D0368988

BOWLING GREEN STATE UNIVERSITY
DISCARDED
LIBRARY

Employment and Labor-Relations Policy

Employment and Labor-Relations Policy

Edited by
Charles Bulmer
John L. Carmichael, Jr.
University of Alabama
in Birmingham

LexingtonBooks
D.C. Heath and Company
Lexington, Massachusetts
Toronto

Library of Congress Cataloging in Publication Data

Main entry under title:

Employment and labor-relations policy.

1. Labor policy–United States–Addresses, essays, lectures.
2. United States–Full employment policies–Addresses, essays, lectures.
3. Discrimination in employment–United States–Addresses, essays, lectures. 4. Industrial safety–United States–Addresses, essays, lectures.
5. Labor laws and legislation–United States–Addresses, essays, lectures.
6. Trade-unions–Government employees–United States–Addresses, essays, lectures. I. Bulmer, Charles. II. Carmichael, John L.
HD8072.E625 331'.0973 79-3145
ISBN 0-669-03388-X

Published simultaneously in Canada.

Printed in the United States of America.

International Standard Book Number: 0-669-03388-X

Library of Congress Catalog Card Number: 79-3145

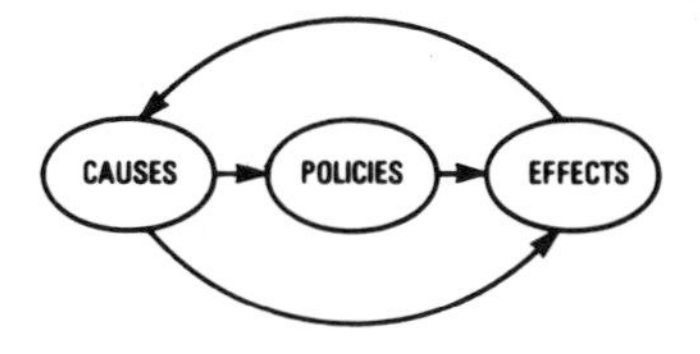

Policy Studies Organization Series

General Approaches to Policy Studies

Policy Studies in America and Elsewhere
edited by Stuart S. Nagel
Policy Studies and the Social Sciences
edited by Stuart S. Nagel
Methodology for Analyzing Public Polices
edited by Frank P. Scioli, Jr., and Thomas J. Cook
Urban Problems and Public Policy
edited by Robert L. Lineberry and and Louis H. Masotti
Problems of Theory in Policy Analysis
edited by Philip M. Gregg
Using Social Research for Public Policy-Making
edited by Carol H. Weiss
Public Administration and Public Policy
edited by H. George Frederickson and Charles Wise
Policy Analysis and Deductive Reasoning
edited by Gordon Tullock and Richard Wagner
Legislative Reform
edited by Leroy N. Rieselbach
Teaching Policy Studies
edited by William D. Coplin
Paths to Political Reform
edited by William J. Crotty
Determinants of Public Policy
edited by Thomas Dye and Virginia Gray
Effective Policy Implementation
edited by Daniel Mazmanian and Paul Sabatier
Taxes and Spending Policy
edited by Warren J. Samuels and Larry L. Wade
Causes and Effects of Inequality in Urban Services
edited by Richard C. Rich
Analyzing Inequality in Urban Services
edited by Richard C. Rich
The Analysis of Policy Impact
edited by John Grumm and Stephen Wasby

Specific Policy Problems

Analyzing Poverty Policy
edited by Dorothy Buckton James
Crime and Criminal Justice
edited by John A. Gardiner and Michael Mulkey

Civil Liberties
edited by Stephen L. Wasby
Foreign Policy Analysis
edited by Richard L. Merritt
Economic Regulatory Policies
edited by James E. Anderson
Political Science and School Politics
edited by Samuel K. Gove and Frederick M. Wirt
Science and Technology Policy
edited by Joseph Haberer
Population Policy Analysis
edited by Michael E. Kraft and Mark Schneider
The New Politics of Food
edited by Don F. Hadwiger and William P. Browne
New Dimensions to Energy Policy
edited by Robert Lawrence
Race, Sex, and Policy Problems
edited by Marian Lief Palley and Michael Preston
American Security Policy and Policy-Making
edited by Robert Harkavy and Edward Kolodziej
Current Issues in Transportation Policy
edited by Alan Altshuler
Security Policies of Developing Countries
edited by Edward Kolodziej and Robert Harkavy
Determinants of Law-Enforcement Policies
edited by Fred A. Meyer, Jr., and Ralph Baker
Evaluating Alternative Law-Enforcement Policies
edited by Ralph Baker and Fred A. Meyer, Jr.
International Energy Policy
edited by Robert M. Lawrence and Martin O. Heisler
Employment and Labor-Relations Policy
edited by Charles Bulmer and John L. Carmichael, Jr.
Housing Policy for the 1980s
edited by Roger Montgomery and Dale Rogers Marshall
Environmental Policy Formation
edited by Dean E. Mann
Environmental Policy Implementation
edited by Dean E. Mann

Contents

Introduction

Considerable interest has been generated by the issues examined in this book. The many diverse problems in the field of labor-management relations have given rise to numerous policy alternatives. Additional substantive analysis of some of these policy alternatives confines the scope and defines the goals of this work.

Specifically, the contributors to the book deal with the problems of full-employment policy, equal employment opportunity, occupational safety and health, labor law reform, and unionization of the public sector. The present undertaking is not intended to be an exhaustive examination of every conceivable labor-management policy area but is intended to be a systematic treatment of certain outstanding issues.

Leon Keyserling, former chairman of the President's Council of Economic Advisers, examines the problem of unemployment and suggests means by which governmental policy can alleviate unemployment. He stresses that full employment is both necessary and feasible during peace as well as war. He maintains that we are not making full use of available resources, and he advocates the use of governmental fiscal policy to promote full employment.

Helen Ginsburg discusses the legislative debate preceding the Full Employment Act of 1946 and subsequent developments culminating in the Humphrey-Hawkins bill of 1978. Harvey Schantz and Richard Schmidt describe the evolution and substance of the Humphrey-Hawkins bill.

An important and much-debated federal program seeking to alleviate unemployment is the Comprehensive Employment Training Act (CETA). Mary Marvel analyzes the social and political impact of certain manpower training programs funded by CETA and presents some surprising conclusions. Another chapter, by Kenneth Tolo, treats a different provision of CETA, that is, the one dealing with the preapprenticeship training program. The potential for greater development of this aspect of the CETA program is considerably and largely untapped.

One more important and very controversial area of employment policy concerns equal employment opportunity. President Carter's recent reorganization of the federal effort to consolidate enforcement procedures is analyzed by Charles Lamb, formerly with the Commission on Civil Rights, who outlines the several policy options available to the president and discusses the alternative selected. Interesting parallels between the *Bakke* and *Weber* cases, based on titles VI and VII of the Civil Rights Act of 1964, are considered by David Gillespie and Michael Mitchell. Michele Hoyman, in her study of the filing of Equal Employment Opportunity Commission (EEOC) complaints, discovers some interesting facts relating to the political characteristics of states and the propensity to initiate actions under the law.

With respect to the improvement of labor laws, questions have been raised concerning the effectiveness of the implementation of the Occupational Safety and Health Administration legislation. In particular, some of the procedures of the agency have been questioned. However, Robert McLean and Ronald Schneck find that leaders of local unions believe current procedures of the Occupational Safety and Health Administration are effective.

One of the important issues in labor-management relations is that of revision of the basic labor law. Charles Bulmer and John Carmichael address themselves to this subject.

Although it was not specifically directed at labor law reform, the Federal Election Campaign Act of 1971 has had an effect on the relative influence of labor and business in the electoral process. Contrary to expectations, Edwin Epstein discovers that provisions of the Federal Election Campaign Act expanded the political influence of business, acting through political-action committees, and consequently reduced the impact of organized labor on the electoral process.

Laura Olson discusses the increasing importance of union pension funds and their potential to serve the interests of both workers and the public. State compliance with workmen's compensation recommendations of the National Commission established under the Occupational Safety and Health Act is the topic examined by Joel Thompson.

The impact of public-employee unions on state and municipal governments is analyzed in three separate chapters. James Seroka discusses some of the structural changes occurring in local government today that are likely to have a direct effect on labor relations for municipal and county governments. The effect of public-sector unionism is the focus of the chapter by Jeffrey Straussman and Robert Rodgers. Russell Smith and William Lyons conclude in their chapter that the evidence does not show a direct correlation between unionization and an increase in municipal wages. Finally, Ezra Krendel and Bernard Samoff examine the controversial subject of unionization of military personnel and suggest approaches to the issue compatible with military requirements and civilian expectations.

The editors and the Policy Studies Organization wish to express their appreciation to the U.S. Department of Labor and especially to Donald A. Nichols, Deputy Assistant Secretary for Economic Policy and Research, for their aid to the symposium on which this book is based. However, no one other than the individual authors is responsible for the ideas advocated here. We would also like to thank Stuart Nagel, coordinator of the *Policy Studies Journal,* for his invaluable advice and assistance in organizing the symposium.

Part I
Full Employment Policy

1 The Problem of High Unemployment

Leon H. Keyserling

Full Employment in Peacetime and Wartime

A discussion of the unemployment problem must commence with the author's definition of what percentage rate of unemployment is compatible with full employment. I use 2.9 percent. Unemployment was reduced to 2.9 percent in 1953. It fell to 1 percent during World War II, but under conditions not relevant now. Both instances were concurrent with price stability. However, I do not imply that price stability should be made a prerequisite to full employment. Full employment is desirable whether or not accompanied by price stability, for reasons I shall discuss.[1]

Full employment does not result automatically from war; unemployment rose to unacceptably high levels when the Vietnam war was in full swing. Nor does war per se produce a single job. Full employment during some wars occurred because we then summoned up the economic and moral awareness that human beings at work are of more value to the economy than when they are idle; unfortunately, we have not recognized this truth in relative peacetime. Further, it is easier to achieve and maintain full employment during peace than under the dislocating conditions of war.

The Manifold Values of Full Employment

There is no merit in the oft-asserted claim that massive unemployment is greatly hurtful to the unemployed alone, and that unemployment is not as hurtful as it used to be because of ameliorating public policies directed toward assistance of the unemployed. At best, the unemployed are far worse off in terms of economic deprivation, human anxiety, and potentials for social and civil discontent and disturbance than when they are employed. And we should get over the notion widely held that jobs are of value only because of what they do for those who hold them. High unemployment is immensely costly to almost all. Jobs exist to produce useful goods and services; the *unemployed do not produce these.*

The full and varied values of full employment were best stated in the final Economic Report of President Truman, issued in January 1953. Therein, this great president said:

> Under the Employment Act, full employment means more than jobs. It means full utilization of our national resources, our technology and science, our farms and factories, our business brains, and our labor skills. The concept of full employment values ends as well as means; it values leisure as well as work; it values self-development as well as dedication to a common purpose; it values individual initiative as well as group cooperation. In the broadest sense, full employment means maximum opportunity under the American system of responsible freedom.
>
> And it is a concept which must grow as our capacities grow. Full employment tomorrow is something different from full employment today. The growth of opportunity, with a growing population and an expanding technology, requires a constantly expanding economy. This is needed to abolish poverty and to remove insecurity from substantial portions of our population. It offers the prospect of transforming class or group conflict into cooperation and mutual trust, because the achievement of more for all reduces the struggle of some to get more at the expense of others.

Costs of High and Secularly Rising Unemployment

During 1953–1978 inclusive, our average annual real economic growth rate was only 3.3 percent, due to five periods of stagnation and then recession, with secularly rising unemployment as the trough of each recession and the peak of each recovery have tended to leave us with higher rates of unemployment than the previous troughs and peaks. For this period, I estimate conservatively that a 4.4 percent average growth rate was needed for sustained full employment, and many other economists have estimated similarly. This takes into consideration the advance in technology, growth in the civilian labor force, and average annual growth rate in productivity in the private economy under conditions of full employment, all much higher than under conditions of repeated stagnations and repeated recessions. To illustrate, the average annual growth rate in productivity in the U.S. private economy was only 2.6 percent during 1953–1960 when the average annual real economic growth rate in real terms was only 2.5 percent; the respective averages were 3.8 percent and 4.9 percent during 1960–1966. Subsequent developments through 1978 repeated, broadly speaking, these relationships.

The tremendous costs of the gross aberrations from full employment are made transparently clear by the following quantifications, derived by methods which I have just described. During 1953–1978, the difference between a full economic performance and the actual performance came to almost $6 trillion (1977 dollars) worth of total national production, and a difference of more than 76 million man-, woman-, and teenager-years of unemployment. The gross national product (GNP) forfeitures resulted, at existing tax rates, in forfeitures by federal, state, and local governments of at least $1.5 trillion in revenue collections.[2] With these additional funds, substantial payments could have been made on the national debt (if they were deemed to be desirable), the total federal deficits of about $231 billion during this period could have been avoided, and the huge balance could have been used to serve well rather than seriously underserve the great domestic national priorities which depend in part or in whole upon public outlays. We would have been a nation of "what we can do" instead of a nation of "what we cannot afford." And manifestly, the financial crises of New York and Cleveland and many other urban areas have been attributable 90 percent or so to the effects of the poor economic performance, resulting in added welfare costs and diminished tax collections, and only 10 percent or so to real or assumed errors in municipal management.

Excuses for High Unemployment

A number of invalid and sometimes frivolous excuses have been offered for the high and secularly rising level of unemployment. To reiterate for emphasis, the recessions since 1953 have tended to register at their trough a higher level of unemployment than the trough of the previous recessions, and most of the recoveries at their peak have tended to register a higher level of unemployment than the peaks of the previous recoveries. Indeed, unemployment at 6 percent in 1978, probably near the peak of the recovery period, was higher than at the trough of three of the previous recessions since 1953.

The first excuse is that the changing structure of the civilian labor force has made it harder to achieve full employment. This is uncomfortably reminiscent of 1939, when there were about 8 million unemployed. It was then said that most of the unemployed were too old, too young, too black, too female, or too unmotivated to be gainfully employed. They would not cross the street to get a job, and they preferred to be on unemployment insurance or relief. But during World War II and some other periods, and without any compulsory manpower programs, the people suffering from one or more of these "disabilities" marched into the factories, and so did farm people, and they performed well.

The attribution of so much of the excessive unemployment to the particular characteristics of the unemployed fails to distinguish between the *causes* of unemployment and the *selection* of those to be unemployed when the overall climate and the deficiencies in national and private policies produce excessive unemployment. It is probably true that a reasonably efficient industrial process selects first for unemployment those who are somewhat less skilled and competent, largely because of earlier disadvantages imposed upon them, than those who are retained on the job. But this is not the basic cause for their unemployment. Unemployment did not rise from about 4 million in 1973 to almost 8 million in 1975 because of such rapid and large deterioration in the quality of the civilian labor force. The selection of the men for drowning and the women and children for saving when the *Titanic* sank was also due to the differences in their personal characteristics. But the number of people who drowned—the cause of their drowning—was that the boat sank and there were only half enough lifeboats to go around. If there had been enough lifeboats, none would have drowned. All would have been saved. So it is with the number of available jobs and the number and characteristics of the unemployed.

Then it is frequently insisted, even today, that unemployment has become less serious because of the characteristics of a large portion of the unemployed. It is indeed one of the most serious and dangerous evils of unemployment that in 1977 the unemployment rate was 4.2 percent among men and 6 percent among women; and that it was 13.9 percent among white teenagers and 36 percent among black teenagers, with the unemployment among black teenagers near the peak of the recovery more than twice as high than it had been during the recession in 1954. But it is *not* true that high unemployment is less intolerable because it concentrates so highly upon selected groups. As to teenagers, white and black, they are driven toward social aberration and crime; moreover, they will soon become adults, and what kind of adults will they be if they experience economic and social rejection when they are teenagers and react accordingly?

The most commonly advanced argument is that unemployed women need jobs less than unemployed men. But women are now more than 40 percent of the civilian labor force, and the percentage will rise. The great preponderance of the women who seek or hold jobs do so because they (1) have no husbands to support them, or (2) are in families in the poverty cellar or living in deprivation (above poverty, but short of an acceptable basic standard of living, BLS, nonetheless) when the husband is unemployed or even when he is employed, or (3) want to help lift their families to still higher levels of income, by no means affluent, so that they may enjoy some of the goods and services which others have and provide reasonable advantages for their children. For simple reasons of human decency, the woman who is able and wants to work should not be denied that opportunity. And in purely economic terms, we must quickly overcome the notion that a growing civilian labor force is a liability rather than an asset. We have tremendous gaps in serving some national needs, such as health care, mass transport, environmental improvement, and urban renewal, which millions of additional jobs can help to close. In this sense, we are still an underdeveloped country.

Bad Unemployment Record

I now come to the very heart of the matter. Despite the chronically poor economic performance since 1953, and especially since 1969, neither the Economic

Reports of the president nor the offerings of economists in general have made any satisfactory attempt, and certainly have provided no satisfactory answers, as to the *reasons* for the poor performance as a first step toward effective solutions.

A very common stance is that unemployment results from inflation. I shall say more later on in this discussion about the relationship between inflation and the business cycle. Suffice it to say at this point that we had a remarkably stable price level during 1922-1929, except for falling farm prices, and that this did not work against the economic imbalances which ended up in the Great Depression. Profits and investment in expansion of the capability to produce grew rapidly, while the wages of two-thirds of all consumers lagged in that they did not keep up with productivity gains, farm income fell, and public outlays grew too slowly. When ultimate demand (consumer spending plus public outlays) lagged too far behind the ability to produce, the Great Crash resulted, and the spark of the stock market could not have produced a conflagration if the dried-out timber were not there. We had falling prices during the Great Depression, but what good did it do that the purchasing power of the dollar was so high when wheat could be bought for 25 cents a bushel and labor obtained for $2 a week? We had a very stable price level before the recession of 1953-1954 and a quite stable price level before the recession of 1957-1958. Price trends are indeed important. But theory and the empirical evidence reveal that both the United States and other countries can have high or low levels of unemployment and production with a rising, stable, or falling price level. Prices are but one tool among many used in the process of producing the end results of real wealth and well-being. The combined use of many tools determines the balanced or unbalanced relationship which govern the end results.

As indicated above with respect to the Great Depression, responsible analysis of the business cycle, applying equally to the most recent quarter century, shows that the main causal factor is a nonsustainable relationship between private investment in plant and equipment which add to production capabilities and ultimate demand in the form of consumer and public outlays combined. During each period of upturn or boom, such as during the period from the first half of 1961 to the first half of 1966, the real growth rate in such investment was two or more times as high as the real growth rate in ultimate demand. From fourth quarter 1977 to fourth quarter 1978, the real growth rate in such investment was 6.1 percent, while the real growth rate in ultimate demand was only 3 percent. Such relationships are not sustainable. And these imbalances have been fed by relative income trends. From first half 1961 to first half 1966, the average annual real growth rate in corporate profits (including inventory valuation adjustment) was 10.3 percent, while the figure for wages and salaries was only 5.8 percent. From fourth quarter 1977 to fourth quarter 1978, the respective figures were 10 percent and 2.8 percent. Whenever these imbalances result in so-called overcapacity, investment is cut back very sharply, and this combined with the more enduring and larger deficiencies in ultimate demand bring on the periods of stagnation and recession. It is no answer to this analysis to point out that investment, being the most volatile element, declines more rapidly than other sectors when these conditions develop, though it had not yet commenced to do so in early 1979. In fact, in early 1979 investment was advancing several times as rapidly as ultimate demand, accompanied by forecasts of a bleak outlook even by those who did not appreciate the basic causes.

Realistic analysis of resource allocation and income distribution is essential whether or not the conclusions reached are the same as mine. But in abjuring such analysis, national policymakers and others have advocated and brought into effect decisions which have accentuated the imbalances. Federal budget policy from 1961 forward has provided in ratio to GNP more for the military (which I am not competent to judge) but much less for domestic priority outlays. Tight money and rising interest rates have been economically disequilibrating and socially regressive. The wage-price guidelines under Kennedy-Johnson and the controls under Nixon were excessively harsh on labor and relatively ineffectual as to prices. In 1964 and several times through 1978, huge tax reductions gave far too much to investors relative to the concessions made to consumers, and distributed the personal tax cuts regressively, not progressively.

Year after year, I predicted that these policies would lead to a slowdown in real economic growth followed by recession, and lead also to more and more inflation. I also forecast that the misdirected tax bonanzas to investors, responsive to cries of "a capital shortage" which were in the main incorrect, would lead to relatively excessive U.S. private investment overseas, consequent exacerbation of our balance-of-payments and goods-and-services deficits, and a strengthening at our expense of the relative competitive position of some other economies, accompanied by a weakening of the dollar. My warnings went unheeded, but all came to pass just as I had foreseen.

Failure of the Trade-off

The dominant and most widely heralded and accepted national economic policy for a decade or longer has been the trade-off, the attempt to restrain inflation by deliberate lifts in unemployment (and slowdowns in real economic growth), despite decades of empirical evidence against this approach. During 1947-1953, unemployment had averaged 4 percent and been reduced from 3.9 percent in the first year to 2.9 percent in the last. Inflation measured by the consumer price index (CPI) had been reduced from 7.8 percent in the first year (occasioned by premature termination of controls and release for spending of so-called savings, which had no production assets to back them up because they had been blown up in explosives through financing of about half of the World War II costs through borrowing) to 0.8 percent in the last year. During 1953-1961, unemployment averaged 5.1 percent and rose from 2.9 percent in the first year to 6.7 percent in the last. The average annual inflation rate was very low at 1.4 percent, but in the last year it was 50 percent higher than in the first, that is, 50 percent higher than when unemployment had been 2.9 percent in 1953. During 1961-1966, the real rate of economic growth averaged annually 5.4 percent; unemployment, while averaging 5.2 percent due to the extraordinarily high rate in 1961, was reduced from 6.7 percent in the first year to 3.8 percent in the last year. Inflation averaged only 1.6 percent, and rose only from 1.2 percent in the first year to 2.9 percent in the last. During 1966-1969, unemployment averaged 3.7 percent, and fell from 3.8 percent in the first year to 3.5 percent in the last. The average for inflation soared to 4.1 percent, rising from 2.9 percent in the first year to 5.4 percent in the last. The basic reason for the soaring inflation was not the slight further reduction in unemployment, but rather the failure to increase taxes promptly and sufficiently when expenditures due to the Vietnam war mounted tremendously.

During 1969-1978, unemployment averaged 6 percent, and rose from 3.5 percent in the first year to 6 percent in the last. Inflation averaged 6.6 percent, and rose from 5.4 percent in the first year to 7.7 percent in the last. In early 1979, with continued application of the trade-off theory, inflation soared beyond the anticipations of any competent forecasters; unemployment remained very high, and was expected to increase.

Long-Range Federal Budget Policy

I have constructed models for optimum economic performance over the years, and redesigned them from year to year in the light of actual experience. These models, on both product and income sides, have set forth the balanced requirements for sustained reasonably full resource use, with due attention to equitable considerations and the great priorities of our needs, domestic and international. The models indicate that, during 1953-1978, federal, state, and local outlays for goods and services were more than $1,535 billion (1977 dollars) below the needed amounts. In 1978 alone the shortfall was more than $122 billion.

The preponderance of these deficiencies in public outlays have been attributable to defaults in the federal budget. The states and localities during the quarter century have increased their debts and enlarged their spending much more rapidly than the federal government, despite the federal government's far greater and more flexible power to increase revenues, and despite the fact that, in almost all areas of the great priorities served by public outlays, there has been recognition of the need to shift toward a much larger assumption of relative responsibility by the federal government. Welfare is but one excellent example of many.

Quite apart from the economic inanity of balancing the federal budget at the expense of the national economy and the well-being of the people, the effort is doomed to failure because the blood of adequate federal revenues cannot be squeezed from the turnip of a repressed economy. Almost the entire cause of the huge and rising federal deficits has been the rising deficiencies in total national production and employment measured against our potentials. During 1947-1953, when we averaged closer to reasonably full resource use than at any time subsequently and got unemployment down to 2.9 percent (and inflation down to 0.8 percent), there was an average annual *surplus* in the federal budget, despite the high costs of the Korean war. Nothing like this has happened since. During 1969-1978, with a real economic growth rate averaging only 2.8 percent and unemployment rising to 6 percent (and inflation to 7.7 percent), the federal budget *deficit* averaged $27.7 billion and was $46.9 billion by 1978.

Looking ahead, I estimate that, in fiscal 1980 dollars, the annual average federal budget outlays during the fiscal years 1980-1983 inclusive should be $42 billion higher than what would result from continuation of current and projected national economic policies and programs as indicated in the president's 1979 Economic Report and Budget Message. But I have also estimated that GNP, measured in fiscal 1980 dollars, during the calendar years 1979-1983 would, in response to policies designed to restore and then maintain reasonably full resource use, average annually $195 billion higher than the results which I estimate would result from policies and programs set forth for now and the years ahead in the two official documents just referred to. With the average annual differences in GNP being so much greater than the average annual differences in total federal outlays, I find that the differences in federal revenues would have these results: under the federal budget outlays projected in the Economic Report and Budget Message and the GNP consequences as I estimate them, the annual average federal budget deficit during fiscal years 1980-1983 would be $25.6 billion, and would be $15.8 billion in fiscal 1983 and $13.6 billion in calendar 1983. Under the GNP projections and federal budget outlays which I set forth as representing the desirable alternative, the figures would be $14.3 billion as the average deficit for the four years, with a balanced budget in fiscal 1983 and a $2.4 billion surplus in calendar 1983—even with much larger budget outlays and their reordering toward greater accent upon the great domestic priorities which require federal monies through direct programs or financial aid to state, local, or private efforts.

Tight Money and Soaring Interest Rates

In general since 1953, the average annual growth in the nonfederally held money supply has been far less than needed for optimum real economic growth, and in the judgment of most economists has been a major factor in the repeated periods of stagnation and recession which have produced the tremendous deficiencies in

GNP and employment opportunity. Comments by some to the effect that at times the money supply has been growing too rapidly have not taken proper account of the general rate of inflation. To illustrate, if the needed real economic growth is 5 percent and the general inflation rate is 6 percent, an adequate growth rate in the money supply would be somewhere in the neighborhood of 11 percent. To say that this growth rate in the money supply would escalate the inflation puts the cart before the horse. If the other policies which I have cited were used to stimulate rather than prevent real optimum growth, this, along with some other measures which I shall mention toward holding prices in check, would bring about a reasonably stable price level, and then the needed average annual growth rate in the money supply would be only about 5 percent.

While the evil consequences of recurrent "credit crunches" have provoked much comment, albeit without correction, the economically and socially hurtfully distributable effects of the prevalent monetary policy have gone virtually unnoticed. During 1952–1978, the computed average interest rate on all public and private obligations increased 160 percent, and this imposed interest charges more than $1.5 trillion in excess of what they would have been if interest rates had remained at the 1952 level. As to the federal budget alone, these excess interest payments during the period as a whole aggregated about $152 billion; in 1978 alone, this excess was about $21 billion or nearly *twice* budget outlays for education or housing and community development or manpower (all three closely related to unemployment). These figures tremendously understate the toll of rising interest costs upon the federal budget (as well as upon state and local budgets) because of the public costs of welfare and unemployment insurance and the lowered revenues which result from the roller-coaster economic performance, with the Federal Reserve so often in the driver's seat.

The prevalent monetary policy has done even more damage to the states and localities, and the excess interest cost of about $1.35 trillion imposed upon private borrowers during 1952–1978 has run counter to economic balance and has been profoundly regressive. All this contributes mightily to unemployment.

Moreover, sharply rising interest costs are inflationary per se, most importantly because the stunting of real economic growth which they have done so much to bring about is inflationary rather than anti-inflationary, for reasons already discussed. In addition, sharply rising interest costs are especially inflationary with respect to production and availability of housing, fuel, and food, because the basic producers of these commodities are unusually dependent upon debt capital. In the case of medical care and housing, inflation also results because of the adverse effects upon the health picture of an excessively restrictive federal budget adopted in the name of reducing inflation. Thus a monetary policy which repeatedly increases both unemployment and inflation approaches the inexplicable.

The foregoing errors in national economic analysis and policies have occurred with rhythmic regularity, with variations of only secondary significance, during the past quarter century to date. And the executive branch policymakers in early 1979 undertook a veritable replay of earlier mistakes.

Viewing the president's 1979 Economic Report and Budget Message, the trade-off between unemployment and inflation still preempts the stage. Goals are established deliberately to increase unemployment from 6 percent in 1978 to 6.2 percent in both 1979 and 1980, and to reduce the real rate of economic growth from 4.3 percent in 1978 to an average of 2.7 percent for the next two years and only 3.8 percent for the full five-year period 1978–1983. Apart from national defense, a no-growth or negative-growth policy is established for the federal budget, neglecting the great priorities of our domestic needs. The prevalent monetary policy is encouraged, and a regressive tax policy is accepted even if not originally proposed. The composition of the GNP goals is a facsimile of the unbalanced relationships among business investment growth and ultimate demand which have been the main factor in our recurrent troubles.

However, there is one big difference between what is being tried now and what was tried earlier. The January 1979 Economic Report and Budget Message involve egregious executive branch disobedience of the comprehensive economic and social legislation approved in 1978 by a majority of 104 in the House of Representatives and by a four to one division in the Senate, and then signed by the president with the promise to help implement it vigorously and without delay. The Humphrey-Hawkins Full Employment and Balanced Growth Act of 1978 merely poured into a single vessel the distilled experience which I have attempted to review in this article. Across the whole gamut of policies which I have discussed, this law *mandates* what we ought to do and *prohibits* what the president is now attempting. And the new bifurcation of legislated policy and executive action is not just a "political" problem. It is a prime economic problem, because the corrosion of confidence flowing from honoring the law in the breach alone can have devastating impacts upon the behavior patterns of functioning economic groups and the people's trust in their government.[3]

I cannot herein propose solutions to the political aspects of this problem. But in view of the ever-increasing influence of economists upon national economic policies and the increasingly nonempirical trend in economic research and study, it may be permissible to suggest that a thorough reconstruction in American economic thought and action is long overdue.[4]

Notes

1. All actual data are based preponderantly upon U.S. government publications. Those used are the Economic Reports and Budget Messages of the president; the Monthly Labor Review of the BLS, in the Department of Labor;

and the Surveys of Current Business of the Department of Commerce. Some publications of the Federal Reserve Board are also used.

2. All estimates (as to the past and projected) of potentials, losses, benefits, and needs, whether related to the economy, the federal budget, or monetary policy, are the work of the author, with the help of staff, and generally are to be found in the publications of the Conference on Economic Progress. The conference is a nonprofit tax-exempt foundation of small size engaged in research, publications, testimony, and writing and speaking related to national economic performance, policies, and programs. Since its establishment in 1954, the conference has published more than thirty book-length paperback studies, listings or copies of which are obtainable from the conference at 2610 Upton Street, N.W., Washington, D.C. 20008.

3. See pamphlet entitled *Where Are National Economic Policies Headed? The President's 1979 Economic Report and Budget versus Fulfilling the Humphrey-Hawkins Act of 1978.* This pamphlet, prepared by the author, was issued by the Full Employment Action Council in March 1979 and may be obtained from the author at the same address as the Conference on Economic Progress. A much more comprehensive discussion of similar issues is contained in *Goals for Full Employment and How to Achieve Them under the "Full Employment and Balanced Growth Act of 1978."* This book-length paperback study was issued by Senator Muriel Humphrey and Representative Augustus F. Hawkins in February 1978, and was prepared mainly by the author and may be obtained from him.

4. For one of the author's many discussions of this subject, see "What's Wrong with American Economics," *Challenge* (May-June), 1973.

Full Employment as a Policy Issue

Helen Ginsburg

In the mid-1970s, for the first time in three decades, full employment became a major policy issue. The Humphrey-Hawkins Full Employment and Balanced Growth bill served as focal point of the contemporary debate. And with passage of the act in October 1978, it is likely that the debate will continue and even intensify as the arena shifts to implementation of the act.

This chapter concentrates on two topics that shed light on some underlying issues in the full employment debate: the origin of the legislative debate over full employment, and the anatomy of unemployment since World War II and its relevance to the full employment issue.

Origin of the Legislative Debate

Before the 1970s, the only time the U.S. Congress gave serious consideration to full employment legislation was in the mid-1940s. It is easy to see why full employment became a major issue at that time.

World War II followed right on the heels of the Great Depression, a traumatic event and a turning point in American history.[1] Unemployment had never been so massive or persistent. Soaring from 3 percent in 1929, it peaked at 25 percent in 1933, when some 13 million were jobless—equivalent to about 25 million in 1979. From 1931 to 1940, unemployment averaged 19 percent.[2]

Economic Theory. Economic theory gave little practical guidance to policymakers faced with a collapsing economy, poverty, mass unemployment, conflict, and chaos. Historically, unemployment had been a recurrent feature of all capitalist economies but interpretations of the nature and significance of the phenomenon varied as widely as policy prescriptions.[3]

At the time, neoclassical economics reigned supreme. It stressed the transitory nature of unemployment and denied or minimized the existence of involuntary unemployment. From a core belief in the self-regulating nature of capitalism flowed advocacy of laissez-faire. The government was admonished not to interfere in the economy, even during times of unemployment. The Great Depression shattered the commanding authority of neoclassical theory. With armies of the jobless clamoring for work, laissez-faire was simply out of the question. Even less appealing to policymakers were the implications of orthodox Marxian theory in which depressions and unemployment were considered inherent features of capitalism. Ultimately they would weaken the system and contribute to its collapse. The introduction of John Maynard Keynes's theory in 1936 seemed to fill an obvious policy void. The influential British economist acknowledged the chronic tendency of advanced capitalist economies to generate high levels of unemployment but built a strong case for active government intervention in the economy, especially to increase expenditures and stimulate demand. Given these techniques, Keynes was certain that a private enterprise economy could eliminate unemployment.

New Deal Policies. Policies and pragmatism, more than theory, shaped New Deal policies against unemployment. The New Deal approach was strikingly similar to that of Keynes but antedated his publication in 1936 of *The General Theory of Employment, Interest and Money.*[4] There were direct efforts to create jobs through the many government-sponsored work projects, of which the most notable was the much maligned Work's Project Administration (WPA). Indirect efforts, such as pump-priming, were also used in the attempt to stimulate private-sector employment.[5]

New Deal job programs were never really accepted by conservative adherents of laissez-faire. Opposition was frequent and often directed toward their gandiose scale. Yet in reality these efforts were never massive enough to end unemployment. In 1941, it still averaged 10 percent and only World War II finally ended the depression.

World War II. Unemployment dropped rapidly to a low of 1.2 percent by 1944. Virtually every civilian who wanted work could find a job. At the peak of the war mobilization, nearly 12 million persons were in the armed forces. That, along with the stimulation of demand caused by skyrocketing military expenditures, led to a labor market typified by shortages. It was one that welcomed and even sought many rejects or "unemployables" of an earlier day. Housewives were called out of the kitchen. Many of the elderly, youths, and handicapped persons got jobs and contributed to vital national production.[6] Blacks also made important breakthroughs, though mostly as manual laborers. In sum, World War II was the most extended period of full employment ever experienced by this nation.

Wartime full employment provided fertile soil for the emergence of full employment as a major policy issue. A few years of full employment could not remove the pervasive fear of a postwar recurrence or the widespread feeling that a nation capable of providing jobs for all during wartime could and should do no less during peacetime.

In 1944, while the battles still raged, Franklin D. Roosevelt formulated a Second Bill of Rights. First on this list was "the right for all to . . . a useful and remunerative job." and the Democratic Party platform promised to "guarantee full employment." In that year's election, the issue seemed beyond controversy.

Not only did Thomas Dewey, the Republican presidential nominee, support full employment but he emphasized the government's obligation to intervene in the economy to provide sufficient jobs when necessary.[7]

The Full Employment Bill of 1945. In reality, conflicting interests and philosophies had not vanished. By 1945, the political sentiment for full employment had entered the congressional arena, where liberal Senators introduced the ill-fated Murray-Wagner Full Employment Bill. The bill had the strong support of a wide variety of liberal, labor, social welfare, and religious organizations, ranging from the Congress of Industrial Organizations to the Young Women's Christian Association. But it was just as vehemently opposed by conservative and business organizations such as the National Association of Manufacturers, the Chamber of Commerce, and the American Farm Bureau Federation.[8] The unambiguous goal of the bill was the establishment of a "national policy and program for assuring full employment." This meant that:

> All Americans able to work and seeking work have the right to useful, remunerative, regular and full-time employment, and it is the policy of the United States to assure the existence at all times of sufficient employment opportunities to enable all Americans who have finished their schooling and who do not have full-time responsibilities to freely exercise this right.[9]

The bill spelled out the federal government's role in assuring full employment. The president was to transmit annually to the Congress a National Production and Employment Budget with estimates of the size of the labor force, the total national production required to provide jobs for that labor force, and total production in the absence of special measures by the federal government. If anticipated production was insufficient to provide jobs for the whole labor force, the federal government was to encourage enough additional nonfederal investment and expenditures to stimulate private-sector employment. However, if necessary, federal investment and expenditures were to be used to close any remaining gap to assure full employment. This bill was an effort to apply Keynesian techniques to attain full employment.

The Full Employment Bill passed by 71 to 10 in the more liberal Senate and was then debated in the House. But by that time, the expected postwar depression had not materialized; the first wave of postwar strikes had started, adding fuel to antilabor feeling; and conservative groups, especially business, intensified their campaign against the bill. The more conservative House thus defeated the bill and offered a substitute, which differed substantively from the Senate bill. The Whittington-Taft Employment Act of 1946 is the compromise that emerged from the struggle.[10]

The Employment Act of 1946. Differing markedly from the Full Employment

Bill of 1945, the Employment Act of 1946 rejected the full employment goal and the federal obligation to intervene in the economy to guarantee jobs for all in favor of an obligation to create:

> "conditions under which there will be afforded useful employment opportunities for those able, willing and seeking work, and to promote maximum employment, production and purchasing power."[11]

For full employment, the vaguer goal of "maximum employment" was substituted and the government commitment was limited to "all practical means consistent with the needs and obligations of national policy."[12] Thus employment was no longer the unambiguous priority, and impractical means of attaining it did not have to be utilized. Other changes were consistent with these. Instead of the National Production and Employment Budget and the explicit pledge to use federal resources to assure full employment if the private sector failed, the act simply required an Annual Economic Report of the president and establishment of the Council of Economic Advisers and the Joint Economic Committee of Congress.

These were not mere semantic differences but represented important philosophical differences concerning the employment objective and the role and extent of government intervention in the economy. As the Conference Report of the Congress stated:

> The House substitute declared that . . . the function of Government is to promote and not to assure or guarantee employment. . . . [In the conference agreement] the term "full employment" is rejected, and the term "maximum employment" is the objective . . . the words or terms "full," "guarantee," "assure," "investment," and "expenditures," do not occur in the conference agreement. The goal is maximum or high levels of employment. The emphasis on expenditures and disbursements is omitted from the conference agreement.[13]

The Employment Act of 1946 is a legislative landmark. But its actual objective was to prevent mass unemployment rather than to attain full employment.

Unemployment since World War II

In the post-World War II era, massive depressions have been averted without eliminating recurrent recessions, significant levels of unemployment between recessions, or an unequal distribution of unemployment. This policy has been costly and has contributed to other socioeconomic and political problems, making solutions more difficult.

Postwar Recessions. There have been six postwar recessions. The last one was the most severe since the Great Depression. Unemployment soared to a postwar annual high of 8.5 percent in 1975 and the number of unemployed *averaged* 7.8 million but 21 million persons were jobless at one time or another during the year.[14] Afterward, unemployment fell slowly; it was still 7 percent in 1977. Even before that recession, unemployment was substantial and drifting upward. Between 1946 and 1974 it averaged 4.7 percent but since 1948 it has never dipped below 4 percent except during the Korean and Vietnam wars.[15] During these years, American rates of unemployment were also higher than in many industrial capitalist nations. Between 1960 and 1970, unemployment averaged 4.8 percent in the United States compared to 0.6 percent in Germany, 1.3 percent in Japan, 1.7 percent in Sweden, 2 percent in France, and 3.1 percent in Great Britain.[16] (Foreign rates adjusted to U.S. concepts of measurement).

The Composition of Unemployment. The official rate masks much of the problem. In the third quarter of 1978, unemployment was 6 percent: 6 million persons.[17] But that excludes about 900,000 discouraged workers, persons no longer looking for work because they think they cannot get employment and 3.3 million persons working part time because they cannot get full-time work. The average rate also masks the inequitable distribution of unemployment, a factor which helped to make full employment a nonissue for so long, especially among the more favored groups. An overwhelming majority of the jobless—more than three out of four—are white. However, unemployment disproportionately strikes minorities, as well as women, youths, the poor, the unskilled, and blue-collar workers. Thus, though overall unemployment was 6 percent, it was 5.2 percent for whites compared to 11.8 percent for blacks; for adult men 20 years and over, it was 3.6 percent for whites but 8.5 percent for blacks; for adult women the comparable figures were 5.4 percent and 10.6 percent; for 16- to 19-year-olds the rates are much higher than for adults and the racial difference is much more pronounced—13.7 percent for whites but 34.8 percent for blacks.

The recession of the mid-1970s worsened but did not create the black unemployment problem.[18] Black unemployment has been about double the white rate since 1954 and that ratio has become more unfavorable in recent years. Relative to young whites, the situation of young blacks has deteriorated considerably since 1954. And since 1958, there has never been a year when the black teenage jobless rate has fallen below 25 percent.

Similarly, in recent decades, the unemployment of women has worsened relative to men.[19] And the real unemployment gap between favored and unfavored groups is even larger than official figures suggest, since minorities, women, and youths comprise the bulk of discouraged and involuntary part-time workers.

Unemployment also has a spatial dimension. Many older cities in the Northeast and the Midwest have experienced a substantial loss of jobs to the South and the West, to foreign lands, and to nearby suburbs.[20] The suburbs experienced employment as well as population growth as the middle class fled from inner cities, encouraged by federal housing and transportation policies. Blacks were locked into major cities with slow or declining job growth. The high jobless rates of inner cities in comparison with suburbs partly reflects this situation. Unemployment often varies as much between a city and its suburbs, and also between poor and middle-class or affluent sections within a city, as among regions. There is a regional dimension to unemployment but the snowbelt-sunbelt dichotomy is an oversimplification. States with the lowest jobless rates are not found in the South but in and near the North Central farm belt. The last recession dealt a severe blow to many sunbelt states. In 1975 annual unemployment rose to 10 percent or more in Arizona, California, Florida, and New Mexico. But even in nonrecession years, unevenness prevails and substantial unemployment persists within states and regions experiencing relatively low jobless rates as well as in regions hardest hit by chronic problems.[21]

Costs of Unemployment. Unemployment is a costly policy. In 1975 every 1 percent rise in the jobless rate led to a $16 billion increase in the federal deficit.[22] Tax revenues fall while expenditures on items such as unemployment insurance, food stamps, and welfare rise. A recession affects many state and local governments similarly.

One of the biggest losses is forgone output. Unused plant capacity and idle workers do not produce goods and services. In 1975 alone, the recession cost the nation $200 billion in lost production—nearly $1,000 for every man, woman, and child.[23] High and prolonged unemployment also contributed to recent short-term social security funding problems. Jobless workers do not pay into the social security fund and more had to be paid out to jobless workers forced into premature retirement.

Unemployment is a major factor underlying the urban crisis and the welfare problem.[24] Unless the poor can be provided with steady jobs at decent wages, the need for welfare is not likely to recede much. Federalizing welfare might reduce the financial burden of states and localities. But cities where a substantial underclass faces few decent job prospects will not be viable places to live. Urban crime will continue to thrive wherever minority youth unemployment is at the disaster level.

Perhaps the highest costs are the human ones. Unemployment means more than lost income. It means loss of a sense of self-worth and has been linked to higher incidences of family disintegration, mental and physical disorders, crime, alcoholism, drug addiction, and other social pathologies. Some of the damage is irreversible. In a seminal study for the Joint Economic Committee, Harvey Brenner of Johns Hopkins University found that a 1 percent rise in the jobless rate over a period of six years has been associated with increases of about 37,000 deaths, more than half of them from cardiovascular disorders.[25] Brenner notes that part of the damage, as in the case of deaths from heart attacks, can occur years after the initial trauma. Seen from this perspective, part of the hidden cost of unemployment means years of higher expenditures on police, prisons, hospital beds, welfare, disability and widow's pensions, and the like.

Full Employment as a Contemporary Issue. Full employment, then, is a necessary though not sufficient ingredient for substantial progress over a wide range of problems such as poverty, crime, welfare, and the urban crisis. It is also essential if there is to be equality for women and minorities and if there are to be ample job opportunities for youths and for older and handicapped persons. Unemployment intensifies intergroup conflicts. Women and men, young and old, black and white view each other antagonistically in the fierce battle for scarce jobs. National unity erodes as regions, states, and cities strive to solve their own unemployment problems by enticing jobs from other areas.

Full employment was a dormant issue for three decades. Interest revived in the 1970s when the impetus came from a broad coalition of civil rights, women's, religious, labor, senior citizens', and other organizations that saw the need not just to end recessions.[26] They wanted to end the recession and depression-like conditions that millions of Americans faced even in so-called good times. Unemployment had not been a political issue between recessions because it was heavily concentrated among groups with relatively little status or political power. The coalition sought to replace the policy of maintaining unemployment at politically tolerable levels with genuine full employment. They also hoped to utilize the additional labor to meet America's unmet needs in areas such as health care, education, transportation, and cities.

The Humphrey-Hawkins Full Employment and Balanced Growth Act of 1978 is a step in this direction.[27] It makes full employment a national policy and establishes the right of all Americans able, willing, and seeking to work opportunities for useful paid employment at fair wages. Specific five-year interim unemployment targets are set: 3 percent for those 20 and over and 4 percent for those 16 and over. Longer-range goals include full employment as soon as possible thereafter and the reduction and ultimate removal of the gap between the unemployment rates of teenagers, women, and minorities, and the overall rate. For the first time a process is established for fomulating national economic policies openly and in a comprehensive, coordinated, and consistent basis in the pursuit of full employment.

The president, moreover, is required to develop programs to reach the employment targets of the act. Job creation in the private sector is to be emphasized, as the order of priority shows: (1) expansion of conventional private jobs through general economic and structural policies; (2) expansion of private employment through federal assistance; (3) expansion of regular public employment; and (4) direct creation by the federal government of additional jobs. The latter, the last-resort public jobs, cannot be put into operation until two years after the act's passage and then only after being authorized and funded by Congress. This would occur after a finding by the president, transmitted to Congress, that other means are not yielding enough jobs to meet the unemployment goals and timetables. Policies and programs are to be geared to putting the jobless to work in priority areas of unmet national need. The act also calls for reducing inflation to 3 percent within five years without impeding the employment goal, which is given priority status.

Passage of the act has not made full employment noncontroversial. For example, labor and business have traditionally viewed the issue differently because full employment shifts power to labor, strengthens unions, makes workers less docile, and puts upward pressure on wages. As menial jobs become harder to fill, low-wage employers and consumers of their products and services may even heighten their opposition to full employment.

Similarly, much opposition to full employment is based on fear of inflation. Based on considerable evidence, the Humphrey-Hawkins Act is premised on a rejection of an inevitable inverse relationship between unemployment and inflation and aims to reduce inflation while pursuing full employment. But conflicts will surely arise between those who still wish to fight inflation with recession-inducing techniques and those who favor structural methods.

Full employment policy is bound to raise myriad other questions and to touch on a host of other issues, most of them controversial. How will full employment be achieved? Should there be mandatory wage, price, and profit controls? What should be the mix of private and public job creation? How much should public jobs pay? What kinds of jobs should be created? What should be the relative size of the military and social budgets? Will growth be accompanied by adequate environmental protection? What would be the impact of alternative employment strategies on energy resources? Should the government regulate the location decisions of private firms? Should private employers get subsidies to employ the jobless? Which programs will have the greatest impact on the groups and areas with the worst unemployment problems? The list could go on and on.

Full employment has not yet come to the United States. But attempts to implement the Humphrey-Hawkins Act may make it one of the leading domestic issues of the 1980s.

Notes

1. See, for example, Broadus Mitchell, *Depression Decade: The Economic History of the United States,* vol. 9 (New York: Holt, Rinehart and Winston, 1947).

2. *Employment and Earnings* 21 (April 1975): table A-1, p. 19. All subsequent data through 1974 are from or derived from this source.

3. For a policy-oriented discussion of economic theories see Daniel R. Fusfeld, *The Age of the Economist,* 3d ed. (Glenview, Ill.: Scott, Foresmen and Co., 1977) or E. Ray Canterbery, *The Making of Economics* (Belmont, Calif.: Wadsworth Publishing Co., 1978).

4. John Maynard Keynes, *The General Theory of Employment, Interest and Money* (New York: Harcourt, Brace and World, 1936).

5. See Mitchell, *Depression Decade,* ch. 5.

6. Russell A. Nixon, "The Historical Development of the Conception and Implementation of Full Employment and Economic Policy," in *Public Service Employment: An Analysis of Its History, and Problems and Prospects,* ed. by Alan Gartner, Russell A. Nixon, and Frank Riessman, pp. 21-22.

7. Quotes from Roosevelt and the Democratic Party Platform and the discussion of Dewey are from or based on Stephen Kemp Bailey, *Congress Makes a Law* (New York: Vintage Books, 1960), pp. 41-43. Much of what follows relies heavily on Bailey's authoritative legislative history of the Employment Act of 1946.

8. Ibid., chs. 3-6.

9. S. 380, 79th Cong., 1st sess., sec. 2(b), (1945).

10. Bailey, *Congress Makes a Law,* chs. 7-8.

11. P.L. 304, The Employment Act of 1946, 79th Cong., 2d sess., sec. 2.

12. Ibid.

13. The House Conference Report on S. 380 is reprinted in U.S. Senate, Committee on Labor and Public Welfare, Subcommittee on Employment, Manpower and Poverty, *Comprehensive Manpower Reform, Hearings,* pt. 5, 92d Cong. 2d sess., 1972. Quote from pp. 2503-2504.

14. U.S. Department of Labor, Employment and Training Administration, *Employment and Training Report of the President, 1978* (Washington: U.S. Government Printing Office, 1978), table A-1, p. 179 and table B-18, p. 260.

15. Helen Ginsburg, *Unemployment, Subemployment, and Public Policy* (New York: New York University Center for Studies in Income Maintenance Policy, 1975), p. 6.

16. Ibid., p. 16.

17. All third-quarter figures are from U.S. Department of Labor, Bureau of Labor Statistics, "Labor Force Developments: Third Quarter 1978," Labor Department press release 78-849, October 16, 1978. The figure for involuntary part-time workers is for August 1978 and is from U.S. Department of Labor, Bureau of Labor Statistics, "The Employment Situation: August 1978," press release 78-753, table A-3. Data for blacks refer to blacks and other nonwhites, of whom about 92 percent are blacks. Other minorities such as Hispanics also have much higher than average unemployment rates.

18. Ginsburg, *Unemployment,* pp. 60-68, and National Commission for Manpower Policy, *The Economic Position of Black Americans: 1976,* special report no. 9, July 1976 (Washington, D.C.: National Commission for Manpower Policy, 1976).

19. Ginsburg, *Unemployment,* pp. 73-76, and U.S. Department of Labor, Women's Bureau, *1975 Handbook on Women Workers,* Bulletin 297 (Washington: U.S. Government Printing Office, 1976), pp. 64-72.

20. Some aspects of this decline are discussed in George Sternlieb and James Hughes, "Metropolitan Decline and Inter-Regional Job Shifts," in *The*

Fiscal Crisis of American Cities, ed. by Roger E. Alcaly and David Mermelstein (New York: Vintage Books, 1977) pp. 145-164. See also James Heilbrun, *Urban Economics and Public Policy* (New York: St. Martin's Press, 1974), especially pp. 37-56 for population and job trends.

21. Department of Labor, *Report of the President, 1978,* table D-4, p. 282; table D-8, p. 290.

22. U.S. Senate, Committee on the Budget, *First Concurrent Resolution on the Budget–Fiscal Year 1976, Report to Accompany S. Con. Res. 32,* 94th Cong., 1st sess. (Washington: U.S. Government Printing Office, 1975), p. 114.

23. Ibid., p. 5.

24. For a discussion of the relationship to the urban crisis and welfare, see Ginsburg, *Unemployment,* ch. 5., or "Needed: A National Commitment to Full Employment," *Current History* 65 (August 1973): 74-75, 88.

25. U.S. Congress, Joint Economic Committee, *Achieving the Goals of the Employment Act of 1946–Thirtieth Anniversary Review,* vol. 1, paper 5. *"Estimating the Social Costs of National Economy Policy: Implications for Mental and Physical Health, and Criminal Aggression,* by Harvey Brenner (Washington: U.S. Government Printing Office, 1976), p. 5. This excellent study also contains a review of the literature and a bibliography.

26. For a discussion of the politics of this coalition and on the development of the Humphrey-Hawkins Bill in its early versions, see Helen Ginsburg, "Jobs for All: Congressional Will-o'-the-Wisp," *The Nation,* February 5, 1977, pp. 138-143.

27. P.L. 523, 95th Cong., 2d sess.

Politics and Policy: The Humphrey-Hawkins Story

Harvey L. Schantz and
Richard H. Schmidt

Humphrey-Hawkins, enacted into law as the Full Employment and Balanced Growth Act of 1978, is the most important step the federal government has taken for overall economic coordination since passage of the Employment Act of 1946. The centerpiece of the new law is specific goals for unemployment and inflation. All federal programs and policies are to work toward achieving a 3 percent adult and 4 percent overall jobless rate within five years, and inflation rates of 3 percent by 1983 and 0 percent by 1988.

This chapter outlines the major changes made in Humphrey-Hawkins from its introduction in June 1974 to its enactment in October 1978, the contents of the Full Employment and Balanced Growth Act of 1978, and the politics surrounding its passage and implementation.

Introduction and Background

On June 26, 1978, Augustus F. Hawkins, a black Democratic congressman representing the Watts, Los Angeles, congressional district, received a minute to address the U.S. House about the Equal Opportunity and Full Employment Bill (H.R. 15476) which he and Congressman Henry S. Reuss (Democrat, Wisconsin) had introduced the week before. "Assuring full employment," began Congressman Hawkins, "is the single most important step in the national interest at this time." Furthermore, according to Hawkins, "an authentic full employment policy rejects the narrow, statistical idea of full employment measured in terms of some tolerable level of unemployment—the percentage game—and adopts the more human and socially meaningful concept of personal rights to an opportunity for useful employment at fair rates of compensation."[1]

The failure of the economy to provide adequate employment opportunity for its almost 92 million civilian labor force was the impetus for Hawkins'

This article is adapted from *Planning for Employment,* a full-length study being jointly conducted by the two authors. Sources include interviews with certain principals (many not for attribution), the public record (media, scholarly, and congressional), and working congressional documents (including party, committee, group, and member shources), as well as participant observation (in the U.S. House and Senate). This chapter was co-authored by Richard H. Schmidt in his private capacity. The authors are solely responsible for all errors of fact and interpretation. This chapter extends and revises our earlier article which appeared in the *Policy Studies Journal* 8 (Winter 1979).

action. After holding at an average annual rate of 4.7 percent from 1962 through 1973, the rate of unemployment began to grow precipitously in 1974.[2] From a 5.2 percent unemployment rate in June 1974, the jobless rate rose to 5.8 percent in September, 6.6 percent in November, and 8.2 percent in January 1975. During 1974 unemployment (seasonally adjusted) increased from 4.66 million to 7.53 million workers, the largest single yearly increase in unemployment during the postworld war period.

Although unemployment increased in every sector of the work force during 1974, blacks, youth, and women (groups having, even in the best of circumstances, a relatively high unemployment rate) absorbed a disproportionate share of this increase in joblessness. Minority unemployment, for example, increased from 9.2 to 13.4 percent; the rate of unemployment among black youth rose from 28.7 to 41.1 percent during 1974.[3] Vernon E. Jordan, Jr., executive director of the National Urban League, commented that this situation "amounts to a major depression for black workers."[4]

The bill introduced by Hawkins and Reuss, and its Senate counterpart introduced August 22, 1974, by Senator Hubert H. Humphrey (Democrat, Minnesota), sought to establish in law a personal right for all Americans willing and able to work an opportunity for employment that would be enforceable in court; require the president to submit an annual economic report to Congress with recommendations that would assure full employment; expand the Local Planning Councils under the Comprehensive Employment and Training Act (CETA) to assess community needs and create a reservoir of public service projects as a potential for job references; provide for the delivery of an actual job by a Job Guarantee Office, an organizational unit in the proposed U.S. Full Employment Service; create a Standby Jobs Corps for jobs of last resort if the private sector could not supply adequate employment; and require a full employment society within five years of enactment.[5]

The central idea behind Hawkins' bill—planning and coordinating the federal economic mechanisms to assure a full employment economy—was not novel. Planning for full employment had informed the efforts of Senators James E. Murray (Democrat, Montana) and Robert A. Wagner (Democrat, New York) during the mid-1940s. The resulting Employment Act of 1946 is generally looked upon as a milestone in the economic history of the United States.[6] The act declared that it is the policy of the federal government to "coordinate and utilize all its plans, functions, and resources for the purpose of creating and maintaining, in a manner calculated to foster and promote free competitive enterprise and the general welfare, conditions under which there will be afforded useful employment for those able, willing, and seeking to work, and to promote maximum employment, production, and purchasing power." Subsequent sections of the 1946 Employment Act provide the statutory basis for the annual economic report of the president, the Council of Economic Advisers, and the Joint Economic Committee of Congress.

Many liberal and labor groups were disappointed in the Employment Act of 1946. Not only had the opposition succeeded in deleting the term "full employment" from the title of the act, they had deflected its purpose by requiring the president to equally consider other economic goals. It was partly "to reverse this calamitous outcome and return to the original intent of the Murray-Wagner full employment bill as introduced in 1945," that Hawkins launched his campaign for full employment legislation.[7]

Planning for Employment

On May 21, 1975, Senators Hubert H. Humphrey and Jacob Javits (Republican, New York) introduced S. 1795, the Balanced Growth and Economic Planning Act of 1975. The thrust of this comprehensive national economic planning measure was much wider than the then current version of Humphrey-Hawkins. A key provision of S. 1795 sought to create an Economic Planning Board in the Executive Office of the President to appraise the nation's total fiscal needs and establish an outline of economic goals. These goals were to be submitted by the president to Congress' Joint Economic Committee for review and for eventual approval by a concurrent resolution of the House and Senate.

On June 11 and 12, the Joint Economic Committee, with Chairman Hubert H. Humphrey presiding, held hearings on "National Economic Planning, Balanced Growth, and Full Employment." One of the highlights of these hearings was the forceful argument of Leonard Woodcock, president of the United Auto Workers, as to the need for improved national economic planning mechanisms. Woodcock argued that the Humphrey-Javits bill and Humphrey-Hawkins "are complementary pieces of legislation. . . . The Humphrey-Javits bill deals with the entire issue of long-range national planning. The Humphrey-Hawkins bill makes it clear that the primary goal of any such national planning must be the achievement of full employment."[8] Future versions of the Humphrey-Hawkins legislation provided an amalgam of these two bills: a national economic planning bill which called for full employment as the primary goal.

Inflation versus Unemployment

Economists and Policymakers

The prevailing belief among economists and policymakers, during the 1960s and early 1970s, was that in modern Western economies there is a trade-off between inflation and unemployment. According to this view, reduction of unemployment below a given level (usually cited today as around 5 percent) through monetary and fiscal policies is inflationary. In Western industrial economies, it therefore follows, full employment and price stability are conflicting national goals.[9]

In recent years, however, this Phillips curve trade-off, inflation for unemployment, has been called into question by a number of economists on both theoretical and empirical grounds.[10] The validity of the inflation-unemployment trade-off was an extremely important point of contention throughout the full employment debate.

Advocates of Humphrey-Hawkins attacked the trade-off on empirical, ethical, and theoretical grounds. The strongest of these criticisms was made by Leon H. Keyserling, a member and chairman of President Truman's Council of Economic Advisers. In testimony before the Joint Economic Committee of Congress, for example, Keyserling termed the trade-off a "pernicious doctrine," which lacked empirical validity.[11]

Within a carefully prepared pamphlet of approximately 100 pages, in which Keyserling had a major role, proponents of full employment legislation contended that the actual operation of the American economy does not support the trade-off; that even if it did, the trade-off would be indefensible in moral terms; and that the roots of inflation are to be found in slack utilization of our productive capacity rather than in a full employment society. In short, reliance on the trade-off makes for bad economic policy.[12]

Opponents of the Humphrey-Hawkins bill vigorously defended the validity of the inflation-unemployment trade-off, and stressed the inflationary impact of any attempt to reduce unemployment to minimal levels. In presenting their arguments, opponents of full employment legislation were able to marshal support from the statements of a "veritable who's who of the economics profession."[13] These included John Kenneth Galbraith, Arthur Okun, and Herbert Stein.

More impressively, opponents of Humphrey-Hawkins could point out that their views were supported by well-placed economists in Democratic Washington. Alice Rivlin, director of the Congressional Budget Office, for example, stated on more than one occasion that fiscal and monetary remedies for unemployment tend to accelerate inflation.[14] Most importantly, Charles Schultze, chairman of the Council of Economic Advisers, was known to have doubts that unemployment and inflation could be simultaneously controlled.[15]

Differences between the two sides of the unemployment-inflation trade-off controversy boiled down to whether the Humphrey-Hawkins bill should or should not include specific anti-inflation goals. The core supporters of Humphrey-Hawkins resisted such a provision. In their pamphlet they stressed that such a goal would probably require the government to institute wage and price controls, a program to which they were opposed.[16] Not mentioned in the pamphlet, however, was the fear, expressed in a Full Employment Action Council leaflet, that "the inclusion of a specific, arbitrary, long-term, numerical inflation goal would invite suspension of the unemployment goal whenever inflation increased, regardless of the cause of particular inflationary pressures being experienced."[17] More than likely, this fear was the primary objection to inclusion of specific inflation goals.

Supporters of an inflation goal jumped upon the weak defense given by their opponents in the pamphlet. The final House minority report on Humphrey-Hawkins quotes the pamphlet defense at length. The minority then contends that the arguments given in the pamphlet offer an implicit rejection of the economic reasoning behind the full employment bill. "H.R. 50 rests on the premise that by reducing unemployment *ipso facto,* you reduce inflation, and without necessitating wage and price controls, which H.R. 50 specifically prohibits." Therefore according to the minority report, "to argue that reducing inflation to 3 percent leads to wage-price controls is to argue that there is a tradeoff and that the bill commits the American people to a rate in excess of 3 percent."[18]

The Carter Administration acknowledged a relationship, rather than a tradeoff, between inflation and unemployment. In important Senate hearings during the spring of 1978, Chairman Schultze and Secretary Marshall presented the administration's point of view on the inflation-unemployment relationship. Both economists felt that a 4 percent unemployment level reached through demand regulation alone would most likely be inflationary. The thrust of their testimony was that a noninflationary expansion of the work force required some concentration on pockets of unemployment—structural measures—as a supplement to macropolicies.[19]

The Public

The crucial development period of Humphrey-Hawkins, 1976 through 1978, was a time of rapid economic expansion. This growth in the nation's total output was accompanied by a declining overall rate of unemployment. Inflation, however, was once again beginning to rise.[20]

These economic changes had a large impact on public attitudes toward the relative importance of inflation and unemployment. In 1975 Americans, by 47 to 34 percent, felt that high unemployment was a greater problem for the country than was rising prices. By 1976 more Americans felt that rising prices was the greater problem. And by August 1978, this plurality had reached 58 to 30 percent. Moreover, in 1978 Americans, by 82 to 10 percent, felt that rising prices rather than unemployment presented the more serious personal problem.[21]

Groups characterized by high levels of unemployment, however, continued to be more concerned with jobs. A national survey released by the Department of Labor on September 12, 1978, reported, for example, that in contrast to the whole population "black Americans are significantly less preoccupied with inflation, their concerns focusing more often on the problems of crime and unemployment."[22] The survey also noted that there was a plurality sentiment among Americans that "it is a responsibility of the government to provide jobs to those who want to work."[23]

Overall, the ever-increasing public concern over inflation during 1976-1978 strengthened the position of those advocating specific anti-inflation goals. In early 1976, though, employment was the goal around which most Democratic presidential hopefuls rallied.

Presidential Politics

The 1976 Campaign and Election

As the 1976 presidential campaign got underway, many Democratic party leaders sought to exploit the unemployment problem, and Humphrey-Hawkins was looked to as the appropriate legislative vehicle. Hubert Humphrey and his staff as well as organized labor, however, were very concerned with the cost of some of the specific provisions of Humphrey-Hawkins. Humphrey and other Democratic leaders feared that the bill had the potential of becoming an albatross around the neck of any Democratic party presidential candidate, as the guaranteed annual income had to George McGovern in his 1972 presidential bid. Already the Republican administration had publicly attacked the cost of the bill, estimated at $30 billion to $60 billion annually.

Representatives of organized labor, Humphrey's office, and Hawkins' office met to discuss their concerns. After long and protracted negotiations, a new version of Humphrey-Hawkins–one that could safely be used as a presidential campaign issue–was agreed upon.

On March 12, 1976, Hubert Humphrey and Augustus Hawkins introduced the new version of their bill at a joint news conference. Their announcement was coordinated with a burst of support for Humphrey-Hawkins by black, labor, and liberal groups. The following week, a two-day conference on unemployment, in commemoration of the thirtieth anniversary of the Employment Act of 1946, was held by the Joint Economic Committee.[24]

The version introduced in March 1976, though calling for 3 percent unemployment within four years, did acknowledge the importance of national production and purchasing power as well as the need to curb inflation. In short, it reflected the influence of the Humphrey-Javits planning bill, and the increasing saliency of inflation.

The new version of Humphrey-Hawkins is "a piece of legislation around which candidates can base their campaigns," Hubert Humphrey noted at the March 12 news conference. And indeed, Democratic presidential hopefuls were doing just that. Humphrey had, from the beginning, viewed Humphrey-Hawkins as a means of returning to the national limelight. Representative Morris K. Udall (Democrat, Arizona) and former Senator Fred Harris (Democrat, Oklahoma) had both endorsed the earlier version of Humphrey-Hawkins. Senator Henry M. Jackson (Democrat, Washington) was expected to support the new version of the legislation.[25]

Presidential candidate Jimmy Carter, through his economic adviser Lawrence Klein, other prominent economists, and freshman Democratic House members, indicated that they did not share Humphrey's enthusiasm with the legislation, however. To meet the objections of these parties, extensive revisions were made. The fall 1976 draft (which was reintroduced on January 4, 1977, the first day of the ninety-fifth Congress) reflected greater emphasis upon the goal of curbing inflation and contained specific anti-inflation provisions as well as made clear that the last-resort reservoir of federally funded jobs would consist of low-paying employment so as not to encourage inflation or employee migration from the private to the public sector.[26]

The Carter Administration

During the initial months of the Carter presidency Senator Hubert Humphrey was, in the opinion of many Washington observers, the president's most effective congressional advocate. In the words of one well-placed staffer, Humphrey time after time rescued the president's "legislative chestnuts from the fire." Humphrey understood the difficulties of setting up a new administration and therefore did not press the White House for strong support of Humphrey-Hawkins. By June, however, Humphrey began to push for action. It was to take considerable pressure—from Humphrey, labor, black leaders, and Speaker Thomas P. O'Neill (Democrat, Massachusetts)—and compromise to win the administration's support for Humphrey-Hawkins.

On November 14, 1977, in a joint news release, the offices of Senator Hubert Humphrey and Representative Augustus Hawkins announced that an agreement had been reached with the president over a new version of the bill. The new draft contained "the provision that adult unemployment be reduced to 3 percent and overall unemployment reduced to 4 percent within five years, in lieu of the earlier provision that adult unemployment only be reduced to 3 percent within four years."[27] Also the president was granted a "bailout" clause; he was given the right to request Congress to alter these goals for the third and subsequent years that the bill was in effect.

The president also received more leeway in the bill's anti-inflationary provisions. The new version of the bill did not require the president to maintain inflation at or below the level at enactment, as did the fall 1976 draft. The new version further stressed the expansion of private employment. The creation of reservoirs of last-resort public service jobs was delayed for at least two years after enactment, and required to have a separate authorization. In short, for his endorsement, the president exacted more flexibility in the legislation's goals and greater reliance on private employment.

Congressional Politics

The House

The Carter-endorsed versions of H.R. 50 and S. 50 were introduced on the first day of the second session of the 95th Congress. On February 8, 1978, the Employment Opportunities Subcommittee of the House Education and Labor Committee, chaired by Augustus Hawkins, approved the latest version of Humphrey-Hawkins. The subcommittee had rejected an amendment by Ronald A. Sarasin (Republican, Connecticut) that would have set a goal for reducing inflation to 3 percent by 1983. This amendment was subsequently rejected by the full committee.

Although the liberal labor-oriented Education and Labor Committee quickly disposed of this attempt to establish an inflation goal alongside the unemployment goal, this issue, as we noted, was to become contentious at almost every step of the legislative process. The importance of the inflation goal grew as inflation rekindled, and continued to be the dominant concern of the American public.[28]

The House of Representatives, after four days of debate and consideration of fifty amendments, passed H.R. 50 by a final vote of 257 to 162 on March 16, 1978. The legislative strategy of the opponents—to alter the bill through the addition of specific goals—was partly successful. The House rejected amendments which would have added a goal of reducing inflation to 3 percent by 1983, and one that called for a balanced federal budget by the same year. However, compromise versions of these amendments, offered by leadership Democrats, were approved by the House. The first of these "requires that the third and subsequent (Presidential) Economic report set forth programs to reduce inflation and progress to that end." The other added that it was one of the purposes of the act "to achieve a balanced budget consistent with the achievement of the unemployment goal."[29]

Additionally, a number of other amendments, including one which called for a goal of 100 percent parity for farm products, were approved. Among the amendments rejected were an attempt at mandating the use of a more restrictive definition of unemployment, and an amendment which provided a reduction in personal and corporate income taxes. In all, the large margin of the final vote did not signify the difficulty the House leadership had in maintaining the integrity of the full employment purpose of the bill. That integrity was going to meet a more severe challenge in the Senate.

The Senate

In the Senate the bill was jointly referred to the Committee on Human Resources and the Committee on Banking, Housing, and Urban Affairs. The Human Resources Committee, led by Employment, Poverty, and Migratory Labor Subcommittee Chairman Gaylord Nelson (Democrat, Wisconsin) and ranking Republican Jacob Javits, reported the bill without major changes on April 13, 1978. The Banking, Housing, and Urban Affairs Committee, however, added four substantial admendments before approving the bill on June 28, 1978. Two of these set goals of 0 percent inflation, and a balanced budget by 1983. The other two were Chairman William Proxmire's (Democrat, Wisconsin) amendment which set a goal of limiting the federal budget to a maximum of 20 percent of the gross national product (GNP), and an amendment offered by John Tower (Republican, Texas) which gave the president power to modify the unemployment goal, of 4 percent or less by 1983, in his first economic message subsequent to the bill's passage. The differences between the two committee versions of the bill, particularly the inclusion of the inflation goal by the Banking Committee, made it difficult for the two committees to write up a joint report on the legislation. The joint report, filed on September 6, 1978, followed two months of tough negotiations in which few of the differences were resolved.[30]

By this time, the scheduled October adjournment made the clock a strong ally of the opponents of Humphrey-Hawkins. Congress was confronted with a host of important issues and but a few weeks to deal with them. Already a group of conservative senators, led by Orrin G. Hatch (Republican, Utah) had threatened a filibuster if the legislation did not include an inflation goal. One senator on the Banking Committee confided that the only way Humphrey-Hawkins could pass was if Congress reconvened after the November elections.

Proponents of Humphrey-Hawkins, particularly the Black Congressional Caucus and the Full Employment Action Council, pressed the president and Senate Majority Leader Robert C. Byrd (Democrat, West Virginia) for action. President Carter and Senator Byrd responded. The president once again reaffirmed his support. And Senator Byrd, in an effort to reach a compromise and avoid a filibuster, appointed an ad hoc committee consisting of key proponents and opponents of the legislation. On October 13, 1978, the ad hoc committee came up with a proposal acceptable to all concerned. In the words of Senator Bob Dole (Republican, Kansas), "we agreed to disagree on inflation and spending goals."[31]

The full Senate decided on the spending and inflation goals on October 13, 1978, when it debated, amended, and passed Humphrey-Hawkins. The version of the bill before the Senate, H.R. 50, did not mention spending limits. Senator Proxmire, however, introduced an amendment which called for a goal of reducing the federal budget to 21 percent of the GNP by fiscal 1981, and 20 percent by fiscal 1983. Instead, the Senate approved language introduced by Senator

Edmund Muskie (Democrat, Maine) setting the goal of reducing federal outlays to their "lowest level consistent with national needs."

Later in the day, the Senate dealt with the inflation goal. The version before the Senate called for an inflation goal of 3 percent by 1983, and 0 percent by 1988. This inflation goal, however, was not to "impede" achievement of the unemployment goals. Senators Muskie and Muriel Humphrey (Democrat, Minnesota) offered an amendment calling for a goal of reducing inflation to 3 percent "at the earliest possible date." Their amendment (in what was perhaps the key vote on Humphrey-Hawkins) was defeated by a coalition of Humphrey-Hawkins supporters, who feared a filibuster if the amendment were approved, and opponents of the bill.

The adoption of the specific inflation goals probably spelled the difference between victory and defeat. It averted a threatened filibuster by amendment on the part of conservative Republicans, and thus cleared the way for Senate approval, by a 70 to 19 vote, late in the evening of October 13. A number of conservative Republican senators voted against final passage, rejecting increased federal economic planning, even though they had succeeded in modifying the goals of the legislation.[32]

Several hours before adjournment on October 15, 1978, the House approved in fifteen minutes, by a standing vote of 56 to 14, the Senate-passed version. President Carter signed Humphrey-Hawkins into law at a White House ceremony on October 27, 1978.

The Full Employment and Balance Growth Act of 1978

Humphrey-Hawkins, enacted into law as the Full Employment and Balanced Growth Act of 1978, is a much different mechanism than the bill introduced by Representative Augustus Hawkins in June 1974. It does not create an absolute guarantee of a job for every able-bodied American. Nor does it establish any employment-creating programs. The act does, however, establish in law for the first time a procedure by which basic economic goals are to be proposed, considered, and established by the president and his advisers, Congress, and the Federal Reserve Board.

Specifically, the new act requires the president to submit in his economic report, annual numerical goals for employment, unemployment, production, real income, productivity, and prices during the next five years. Short-term goals are to be established for the first two years, medium-term goals for the following three years. The purpose of these short- and medium-term goals is not only to achieve full employment and production, but a balanced budget, improved trade balance, and federal outlays at the lowest percentage share of the GNP as practical.

The principal elements of the new law are specific goals for unemployment and inflation. All programs and policies implemented under the act are to work toward achieving a 3 percent adult and 4 percent overall jobless rate within five years, and inflation rates of 3 percent by 1983 and 0 percent by 1988.

Although specific inflation goals were added by the Senate, the primacy of the attack on unemployment was preserved by the proviso that "policies and programs for reducing inflation shall be so designed so as not to impede achievement of the goal and timetables" on unemployment.

Equally important, flexibility in goal setting as requested by President Carter was retained. Beginning with the second economic report in January 1980, the president may modify both the 4 percent unemployment goal and the 3 percent inflation goal and timetable where changed economic circumstance require it. In another title of the act Congress is given the opportunity to review, accept, modify, or reject the changed goals. However, the act provides for a continued commitment to the goal of reducing unemployment to 4 percent as soon as possible.

The other instrument of economic policy making, the Federal Reserve Board, is also required to report to Congress. The board must report twice a year (February 20 and July 20) explaining the monetary policies it is pursuing in furtherance of the unemployment and inflation goals.

The legislation stresses that none of its provisions shall be used to control production, employment, allocation of resources, wages, or prices. The act emphasizes that increases in available jobs should come about by growth in the private sector. Only if the private sector proves inadequate can federally subsidized jobs, public employment, and reservoirs of employment projects be used to ameliorate unemployment.

The employment reservoir is a final option. It can only be triggered by the president if he finds that the other procedures listed are not working. It is further restricted by the requirement that any new employment program that involves more than an expansion of existing law must be authorized by Congress and cannot go into effect until after October 27, 1980. In addition, to satisfy the critics who had opposed the employment reservoir because it would be inflationary, this provision also required that any job must be in the low-pay range and should be designed so as not to draw workers from private employment.

To coordinate the branches of government, the act amends the Congressional Budget Act to establish procedures for congressional review of the goals, programs, and policies recommended in the president's annual economic report. The Joint Economic Committee is to act on the president's recommendation and then submit a report to the respective Senate and House Budget Committees. The Budget Committees are to address the goals and timetables in their report on the first concurrent resolution on the budget.

The Debate Continues

Coinciding with the enactment of the Full Employment and Balanced Growth Act of 1978 was a deepening concern within the Carter Administration over inflation. During 1978, the unemployment rate had stabilized at around 5.8 percent. Inflation, however, had peaked to an average annual rate of 11.4 percent during the second quarter of 1978.[33]

On October 24, 1978, three days before signing Humphrey-Hawkins into law, President Carter had announced a new anti-inflationary policy. The president's program included a call for voluntary wage and price guidelines. These asked businesses to limit price increases to 0.5 percent below their average annual rate of increase for 1976-1977, an annual average increase of 5.75 percent. The hoped-for wage limitation was 7 percent annually. Complying workers were to be protected by a wage-insurance system, which the president said he would propose to Congress.

The president also called for greater austerity in government, and so ordered a cutback in federal hiring. The president said he hoped to trim the federal deficit to $30 billion, half of what it was when he took office. Carter, additionally, promised to cut the federal government's share of total national spending from 23 to 21 percent of the GNP by 1980 rather than 1981, as he had promised during the 1976 campaign.[34]

The administration's strong concern with inflation was reflected in the annual economic report of the president, transmitted to Congress January 1979. As required by the Full Employment and Balanced Growth Act of 1978, this document included short- and medium-term projections for employment, unemployment, production, real income, productivity, and prices. Projected for 1979 and 1980 was an unemployment goal of 6.2 percent, a rate 0.4 percent higher than the then current unemployment rate.[35]

The announced economic goals, but especially the unemployment goal, infuriated advocates of a strong full employment policy. During floor consideration of the 1980 Budget Resolution, for example, Augustus Hawkins argued that the administration's plan "violates the intent of the Humphrey-Hawkins Act for 1979 and 1980 and would make it utterly impossible to reach the mid-1983 goals."[36] In a recent workshop on full employment, Leon Keyserling bluntly stated that in making his economic projections, President Carter was "flagrantly avoiding the law."[37]

The administration's response has been that the nation's ability to achieve the goals of the act depends primarily on moderating inflation. Continued high inflation, argues the Carter Administration, makes impossible the economic stability and investment necessary for a full employment economy.[38]

Passage of Humphrey-Hawkins, then, has not resolved, even temporarily, differences over the direction of national economic policies. Although a major piece of goal-setting legislation has been placed in the statute books, the essential economic debate continues.

Notes

1. *Congressional Record,* 93d Congr. 2d sess., June 26, 1974, pp. H 21278–21283.

2. *Employment and Training Report of the President,* transmitted to Congress 1978, p. 210, table A-18.

3. "Employment Status by Age, Sex, and Color, Seasonally Adjusted," *Monthly Labor Review* 98 (March 1975): 88.

4. Quoted in "Continuing High Rates of Minority Joblessness Prompt Calls for Full Employment Legislation, *Congressional Quarterly Weekly Report* (July 19, 1975): 1588–1589.

5. This paragraph is based on information placed in the *Congressional Record* by Representative Hawkins in the pages indicated in note 1. Virtually identical versions of the bill were introduced in the House and Senate at the outset of the 94th Congress, January 1975. These and succeeding versions of Humphrey-Hawkins bore the designations H.R. 50 and S. 50.

6. The legislative history of this law is eloquently told in Stephen Kemp Bailey, *Congress Makes a Law: The Story Behind the Employment Act of 1946* (New York: Vintage Books, 1950).

7. Hawkins, *Congressional Record,* June 26, 1974, p. H 21278.

8. Leonard Woodcock, "Testimony," *National Economic Planning, Balanced Growth, and Full Employment, Hearings before the U.S. Congress, Joint Economic Committee,* June 11 and 12, 1975, pp. 4–8.

9. See Arthur M. Okun, "Conflicting National Goals," in *Jobs for Americans,* ed. Eli Ginzburg (Englewood-Cliffs, N.J.: Prentice-Hall, 1976), pp. 59–84.

10. A survey of theoretical developments is Thomas M. Humphrey, "Some Recent Developments in Phillips Curve Analysis," *Economic Review* 64 (January/February 1978): 15–23. A discussion of the empirical evidence is in Muriel Humphrey and Augustus F. Hawkins, *Goals for Full Employment and How to Achieve Them under the Full Employment and Balanced Growth Act of 1978 (S. 50 and H.R. 50)* (Washington, D.C.: Leon H. Keyserling, February 1978), pp. 35–38, 48, 49. A readily accessible post-Phillips curve analysis of the U.S. economy is Henry C. Wallach, "Stabilization Goals: Balancing Inflation and Unemployment," *American Economic Review* 68 (May 1978): 159–164.

11. Leon H. Keyserling, "Statement," *Thirtieth Anniversary of the Employment Act of 1946–A National Conference on Full Employment, Hearings before the Joint Economic Committee,* March 18 and 19, 1979, pp. 298–303.

12. Humphrey and Hawkins, *Goals for Full Employment,* pp. 34–51. An earlier version of this pamphlet was published in January 1977 by Hubert H. Humphrey and Hawkins.

13. "Minority Views," *House Report 95–895,* p. 37.

14. See Rivlin, "Prepared Statement," *Thirtieth Anniversary,* pp. 276–285, as well as the statements noted in "Minority Views," *House Report 95–895,* p. 37.

15. See Schultze, "Statement," *Hearings on S. 50,* U.S. Congress, Senate, Subcommittee on Employment, Poverty and Migratory Labor of the Committee on Labor and Public Welfare, May 14, 1976, pp. 3-6 (mimeographed); the February 1978 reference noted in "Minority Views," *House Report 95-895,* p. 37, and the statement in "Minority Views," *House Report 95-895,* p. 41.

16. Humphrey and Hawkins, *Goals for Full Employment,* pp. 46-47.

17. The Full Employment Action Council is a Washington-based group organized to promote passage and now working for implementation of a full employment economic policy.

18. "Minority Views," *House Report 95-895,* p. 41.

19. See the testimony of and the colloquy involving Marshall and Schultze in U.S. Congress, Senate, Committee on Banking, Housing and Urban Affairs, *Hearings on S. 50,* May 8, 9, and 10, 1978, pp. 184-211. Also see the *Economic Report of the President,* transmitted to the Congress January 1978, pp. 106-110.

20. Phillip L. Rones and Carol Leon, "Employment and Unemployment during 1978: An Analysis," *Monthly Labor Review* 102 (February 1979): 3-12; Craig Howell et al., "Price Changes in 1978—An Analysis," *Monthly Labor Review* 102 (March 1979): 3-12.

21. The data in this paragraph were released in *The Harris Survey,* March 20, 1978, and August 31, 1978.

22. Albert H. Cantril and Susan Davis Cantril, *Unemployment, Government and the American People* (Washington, D.C.: Department of Labor, Public Research, 1978), p. 20.

23. Cantril and Cantril, *Unemployment,* p. 86.

24. U.S. Congress, *Thirtieth Anniversary of the Employment Act of 1946—A National Conference on Full Employment, Hearings before the Joint Economic Committee,* March 18 and 19, 1976.

25. Martha V. Gottron, "New Bill: Full Employment," *Congressional Quarterly Weekly Report* (March 20, 1976): 641-642; Bob Rankin, "Candidates on the Issues: Full Employment," *Congressional Quarterly Weekly Report* (March 6, 1976): 513-514.

26. *Congressional Quarterly Weekly Report* (August 21, 1976): 2278; U.S. Congress, House, Committee on Education and Labor, *Full Employment and Balanced Growth Act of 1978,* House Report 95-895, February 22, 1978, pp. 6-7.

27. *House Report 95-895,* p. 8.

28. See chapter section, The Public.

29. These provisions are noted in a document printed for the use of the Senate Committee on Banking, Housing, and Urban Affairs, *Agenda: Full Employment and Balanced Growth Act of 1978,* S. 50, June 1978, pp. 7, 5.

30. U.S. Congress, Senate, Committee on Human Resources and Committee on Banking, Housing, and Urban Affairs, *Full Employment and Balanced Growth Act of 1977, S. Report 95-1177.*

31. Quoted in Associated Press, "Senate Snaps Deadlock on Employment Bill, Agrees to Final Vote," *Fort-Lauderdale News and Sun-Sentinel,* October 14, 1978, p. 8A.

32. See the statement by Senator Barry Goldwater just prior to the passage of H.R. 50, in *Congressional Record,* October 13, 1978, p. S 18968.

33. "Unemployment Rates, by Sex and Age, Seasonally Adjusted," *Monthly Labor Review* 102 (February 1979): 79; Craig Howell et al., "Price Changes in 1978," p. 4, table 1.

34. See "President's Anti-Inflation Program, Address to the Nation, October 24, 1978," *Weekly Compilation of Presidential Documents* 14 (October 30, 1978): 1839-1848.

35. *Economic Report of the President,* transmitted to Congress January 1979, p. 109, table 2.2.

36. Quoted in Christopher R. Conte, "Democratic Coalition Nears Budget Test," *Congressional Quarterly Weekly Report,* 37 (May 5, 1979): 815-817.

37. Policy Studies Organization Panel on Labor and Employment Policy held at the Annual Meeting of the American Political Science Association, Washington, D.C., August 31, 1979. A systematic statement of the Hawkins-Keyserling viewpoint is Full Employment Action Council, "Where Are National Economic Policies Headed?" (Washington, D.C.: March 1979).

38. Charles Schultze, Chairman of the Council of Economic Advisers, "Testimony," *Oversight Hearings on the Full Employment and Balanced Growth Act of 1978,* U.S. Congress, House, Subcommittee on Employment Opportunities of the Committee on Education and Labor, February 13 and 14, 1979, pp. 75-93.

4 The Social and Political Consequences of Manpower Training Programs: The Case of CETA

Mary K. Marvel

Governments long have been involved in investing in human capital through the provision of public education and more recently through federally funded manpower training programs. The Comprehensive Employment and Training Act (CETA) of 1973 represents one of the more recent vehicles for human capital investment by the federal government. Justifications for the government's involvement in human-capital investment can be found in classical democratic and human-capital theories. The benefits of an educated citizenry with shared allegiances are said to accrue to the individual citizen as well as manifesting themselves in important systemic consequences. Individuals who possess the necessary skills and opportunity to participate in the mainstream economic and political systems are less likely to be available for mass movements, thus enhancing societal stability. The benefits of a well-educated citizenry possess characteristics of a public good; if they exist for one, they exist for all. As a result, private decisions result in underinvestment in human capital, precipitating government involvement.[1]

In addition to the public-good argument, it is suggested that society may have preferences about the desired distributions of income. The goals of political and social stability could motivate policies to alter the current distribution of income. The government can make direct transfer payments to the poor or initiate human-capital-investment programs. The choice between these two alternatives has generally been made on economic grounds. Even if a strong economic argument cannot be made for investment in certain human-capital programs, the programs may remain the preferred choice. "When the values of self-produced consumption goods (including self-respect) and complementary consumption goods received in the process of working are included a poor financial investment may be a good social investment."[2]

Both the impetus for and objective of manpower training programs lie within the political as well as economic realm. Traditionally, however, impact evaluations of manpower training programs have been the domain of economists. As a result, the bulk of knowledge available details economic benefits of participation in manpower training programs. The specification of noneconomic consequences of program participation is scarce. In a survey of the relevant literature, it was reported that only seventeen of the more than 200 studies reviewed supplied any information concerning program effects other than employment and earnings.[3]

This analysis focuses on selected noneconomic impacts of CETA participation. A variety of noneconomic impacts are explored with primary emphasis accorded work-related attitudes and more overtly political attitudes. The discussion here focuses on commitment to the work ethic, feelings of political efficacy, knowledge, and electoral participation. Section one describes the data and methodology employed. The hypotheses to be examined empirically are specified in the next section. The major results obtained for each noneconomic impact follow. The analysis concludes with a discussion of the implications for manpower policy formulation and evaluation.

Data and Methodology

The purpose of CETA is "to provide job training and employment opportunities for economically disadvantaged, unemployed and underemployed persons, and to assure that training and other services lead to maximum employment opportunities and enhance self-sufficiency."[4]

Title I of the act, the one to be emphasized in this analysis, establishes a program to provide comprehensive manpower services throughout the nation, including the development and creation of job opportunities and the training, education, and other services needed to enable an individual to secure and retain employment at his maximum capacity.

Prime sponsors generally use title I funds for on-the-job training (OJT), classroom or institutional training, work experience, services to clients, and public service employment.[5] The first three activities are the foci for this study. These three program components have been chosen as they comprised about 90 percent of cumulative title I enrollments nationally as of September 1977, with OJT accounting for 12 percent; classroom training, 38 percent; and work experience, 39 percent.

Any external evaluation of a manpower training program cannot conform to the tenets of the classical experimental method. The inability of the outside researcher to assign participants to various programs and withhold services from a control group precludes randomization, the distinguishing characteristic of an experiment. If evaluation mechanisms are not incorporated in the planning process and implemented at the intake level, the controlled experiment is not feasible. The researcher is inevitably confronted with a situation in which assignment to programs has been made on the basis of decision rules established within an organization. These decision rules can encompass a wide range of intake policies from a first-come, first-serve strategy to a "creaming" policy, selecting those with the greatest likelihood of completion. Random assignment will not often characterize a prime sponsor's intake system.

In this analysis the lack of control over participants' program assignment necessitated that an approximation to an experiment be undertaken. From the groups constituted by the four Ohio prime sponsors chosen for inclusion in the study, participants were selected randomly in each of seven groups. A random sample was drawn from the pool of currently enrolled OJT, work-experience, and classroom-training components in each of the four prime sponsors. The same procedure was used for those who successfully completed the program in three title I program components in each of the prime sponsors. Similarly, a random sample was drawn from the pool of eligible applicants who were not admitted into the program.

There has been considerable debate over the appropriateness and method of construction of control groups in manpower evaluation research. A great many studies evaluating the economic impacts of training programs have not utilized control groups. Those which have employed control-group techniques have used a wide variety of methods. Some researchers have used control groups drawn from the target population while others have used the snowball technique, asking trainees to refer others who are unemployed. Some have attempted to construct a control group by matching on theoretically relevant variables. Still others have used dropouts from the program and qualified nonapplicants.

Several control techniques are used in this analysis. Eligible nonenrollees serve a control function. It is important to note that these individuals have been certified as being eligible for title I. This eligibility requirement removes one source of potential bias. As the nonenrollees have demonstrated the initiative to enroll for CETA services, the disparity in motivation between participants and nonenrollees is minimized. Another control technique is the use of explicit comparison among program component groups. The multiplicity of comparisons available using this design enables one to avoid some of the difficulties associated with any single control technique.

The method of data collection used in this study was the mail questionnaire. After a pretest, questionnaires were mailed to 903 individuals in the four prime sponsors. Of those that could be delivered, 57.4 percent completed and returned the questionnaire.[6] The similarities found between respondents and nonrespondents and respondents and undeliverables, in terms of race, sex, educational attainment, average age, economic disadvantagement, and average pre-CETA wage aid in the problem of nonresponse bias. While caution is necessary in the interpretation of the results, the striking similarities displayed by the three categories on selected characteristics contribute to one's confidence that respondents are representative of the sample.[7]

One final point about the sample needs to be made. As a panel design is not employed, the notion of change remains elusive. One method for tentatively assessing change in the current study is to make the assumption that the most significant feature distinguishing program participants in any given program component from program graduates in that same component is program completion. The latter group, program graduates, are assumed to be similar to current participants except for the fact that they have completed the program. This assumption is predicated on the notion that intake policies remain virtually unchanged. The same decision rules used to route individuals to one program component are not expected to have changed measurably from the time graduates and current participants were involved in the intake process. Empirical support for this assumption is derived from an examination of the characteristics of current program participants and graduates.

A wide range of analytical techniques is employed in this study. The first task is to assess the distribution, variability, and central tendencies of the variables under investigation. Next the relationships among variables are examined. Cross-tabulations—contingency table analysis—are the initial mechanism to determine joint frequency distributions. Bivariate correlation analysis permits a summary of the relationship between two variables. Multiple regression techniques enable one to identify the independent impact, holding other variables constant, of certain theoretically relevant variables on the dependent variables.

Results are aggregated by program component across prime sponsors. Prime sponsor-specific results are investigated to preclude the masking of any important differences through aggregation. For some purposes all current participants and all graduates are grouped together for analysis.

Hypotheses

It is widely recognized that considerations apart from those of economic efficiency play an important role in the formation of social action programs. Both classical democratic theory and human-capital theory provide insights into the nature of the social and political factors that provide an impetus for government-funded manpower training programs. However, unanimity does not prevail as to the means by which disadvantaged members of society should be aided by government programs. The root of the controversy lies in different philosophies concerning the nature of poverty. At least two theories that have influenced policy decisions can be identified. The culture-of-poverty argument identified with Oscar Lewis and Michael Harrington suggests that a significant number of the poor have rejected core societal values and constitute a contraculture. Proponents of the situational approach to poverty, rejecting the subculture argument, minimize the impact that personality disorders and class-specific child-rearing practices have upon the goals of the poor. The argument is advanced that the goals of the poor are substantially the same as those of the middle class, but the reward structure faced by the poor dampens their aspirations. These two views yield contrasting implications for the design of manpower policy.

CETA (title I) is capable of reflecting elements of both the culture-of-poverty and situational arguments as the local prime sponsors enjoy a great deal of flexibility concerning the content of the title I components. The relevant components for this analysis are OJT, classroom training, and work experience. One cannot unequivocally classify any one of the programs according to philosophy. It is possible, however, to make some generalizations bearing in mind that exceptions exist. In general, OJT epitomizes the situational theory as participants are placed in real-world work environments in the private sector. Other CETA services such as counseling and transportation may have been provided to the OJT participant as well, but the thrust of the program is skill acquisition in the private sector. Similarly, classroom training is likely to emphasize the development of marketable skills. Work experience, on the other hand, is a "short-term and/or part-time work assignment with a public employing agency and is designed to enhance the employability of individuals who have either never worked or who recently have not been in the competitive labor population for an extended period of time. The work experience activity is designed to increase the employability of such individuals by providing them with experience on a job, an opportunity to develop occupational skills and good work habits and an opportunity to develop specific occupational goals through exposure to various occupational opportunities."[8] Much more than in either OJT or classroom training, the development of good work habits is emphasized in work experience rather than specific skill development. Accordingly, it is more easily justified in culture-of-poverty terms.

A portion of Leonard Goodwin's study on work orientations is replicated; specifically, commitment to the work ethic.[9] The expectations associated with the relationship between the work ethic and the culture-of-poverty and situational arguments are as follows:

> 1. One might expect that participants exposed to skill training (OJT and classroom training) would evince greater commitment to the work ethic because the reward structure they will be facing after program participation has been improved by the training. The situational approach leads one to expect the following rank ordering among respondents in terms of commitment to the work ethic:
>
> Highest Commitment
> 1. OJT
>   Classroom Training
> 2. Work Experience
> 3. Eligible Nonenrollees
> Lowest Commitment
>
> The culture-of-poverty argument results in the following expectations:
>
> 1. Program participants will score lower on the work ethic than the general population. (Although they may still score higher than the poor generally.)
>
> 2. The group most likely to post a higher score compared to eligible nonenrollees would be work-experience participants. This follows from the emphasis placed on the development of good work habits in the program.

If a different program ranking from that outlined for the competing hypotheses were obtained, an a priori judgment about the cause of the ranking would be difficult. It could be a failure in the theory underlying the program, failure of implementation of the program, or failure in the measurement of the program. It is possible that neither approach provides an adequate framework on which to construct a manpower training program. It is also possible that the theory which motivated the program is sound but the program has been poorly implemented. It has failed to attain its goals, both stated and intuited. Only after considering the totality of the evidence could one make a tentative judgment.

Two competing hypotheses concerning the longevity of any noneconomic benefits of program participation are also examined empirically. One line of argument holds that the effects of the program, especially noneconomic ones, may not show up until after program participation. It has been suggested that these benefits may accumulate over time and be reinforced by postprogram experiences.

Another view of noneconomic effects holds that they will be strongest during program participation and will deteriorate over time. It can be argued that positive feelings are at a zenith during enrollment as the possibility of new jobs and opportunities loom ahead. Disillusionment sets in upon completion of the program when reality does not measure up to expectations. Tentative support for the "cumulative" model, with impacts increasing over time, would be derived from the following ranking of groups: graduates, trainees, and comparison group. In this instance graduates would have exhibited greater adherence to middle-class values under investigation than did current participants. The comparison group would score lowest in this respect. The following ranking would lend tentative support to the "transitory" model: trainees, graduates, and comparison group.

Results

Table 4–1 reports the mean scores and standard deviations for all 440 respondents as well as for respondents in each of the seven groups. It is interesting to note that eligible nonenrollees, those who did *not* participate in any CETA program, evinced a stronger commitment to the work ethic than did any program participant or graduate. The interpretation to be accorded this finding is clouded by the fact that a panel design was not a possibility for this study thereby making the assessment of change tenuous. However, to the extent that current and graduate groups of any given program component are similar with regard to theoretically relevant characteristics, tentative conclusions regarding the notion of change can be offered by comparison of scores for the two groups.

The matter of intake procedures and priorities is critical in any explanation of the work-ethic results. One can envision several scenarios. One can assume a conscious decision on the part of the CETA staff to admit to the program those individuals with the least likelihood of obtaining employment on their own. One indicator of this could be the interviewer's perception of the applicant's avowed

Table 4–1
Mean Work-Ethic Scores for All Groups Included in the Study

Group	*Work-Ethic Mean*
All respondents (N = 440)	3.10 (0.59)
Current OJT participants (N = 43)	3.18 (0.44)
OJT graduates (N = 38)	3.15 (0.53)
Current classroom training participants (CRT) (N = 68)	3.15 (0.47)
Classroom training graduates (CRT) (N = 80)	3.10 (0.56)
Current work experience (WE) participants (N = 86)	3.13 (0.69)
Work experience (WE) graduates (N = 50)	2.65 (0.73)
Eligible nonenrollees (N = 75)	3.24 (0.48)

Notes: Range of values is from 1.0 to 4.0 with 1.0 representing the least commitment and 4.0 representing the highest commitment to the work ethic.

Standard deviations are in parentheses.

commitment to work. A policy decision not to allocate resources to those individuals whose perceived attitudes indicate that they would be able to cope in the world of work but rather to utilize CETA funds on those who have more ambivalent attitudes toward work could serve to explain the lower work-ethic scores for program participants than eligible nonenrollees. To the extent that program completion is the fact distinguishing graduates from their counterparts currently enrolled in the same program component, the even lower scores posted by the graduates would seem to undermine the notion that program completion can heighten the commitment to work.

In the above scenario, program participants were depicted as being less committed to the work ethic than nonenrollees when they were granted admittance to the program. Another equally plausible scenario allows for the possibility of creaming, selecting those judged most likely to successfully complete the program. Creaming on one dimension of motivation, commitment to the work ethic, is possible. Program participants, by definition, would have higher work-ethic scores than those not admitted to the program. If this selection process were in effect, the measured scores for both participants and graduates compared to eligible nonenrollees would be an understatement of the true differences between nonenrollees and participants. Similarly, the stated differences between the two groups under the previously described selection process would be an overstatement of the true differences. Regardless of the intake assumptions made, the factor of critical importance at this time is the inability of participants and graduates to equal the work-ethic scores of the eligible nonenrollees both during and after program participation.

A comparison of the findings of this analysis and those obtained by Goodwin is instructive. He found virtually no difference in the commitment to the work ethic between outer-city middle-class blacks and whites and short- and long-term welfare recipients. It is interesting to note that the scores posted for each of Goodwin's groups are higher, indicating a stronger commitment to the work ethic, than those in this study. One possible explanation lies in the differences in the mode of data collection utilized by the two studies. While Goodwin used a mail questionnaire, the majority of his data was derived from personal interviews; in this study mail questionnaires were the predominant method. Despite safeguards reported by Goodwin, it is possible that an overstatement of work-ethic commitment was obtained because of the potential for the respondents to give the perceived socially desirable response. Of course, the time elapsed between the two studies, ten years in the case of some of the earliest interviews conducted by Goodwin, might be a factor in explaining the differences.

The results of cross-tabulation between the work ethic with race, sex, and economically disadvantaged status for all respondents did not reveal relationships among the variables. To test for the possibility that aggregation across program components may have masked relationships existing within each program category, the same cross-tabulations were performed for each of the seven groups. The same results, that is, no relationships, were found.

Several other bivariate relationships between the work ethic and certain theoretically relevant variables including level of educational attainment, age, and pre-CETA wage were examined. The expected relationship between the the work ethic and education is positive. All groups, except three, exhibit this relationship. While the negative correlations for classroom training and eligible nonenrollees are not statistically significant, that for work-experience graduates is of statistical and perhaps substantive importance. It is possible that expectations raised during program participation are soured by the graduates' inability to locate meaningful employment. This disappointment might be felt most keenly by those with higher-education levels.

The expected relationship between age and work ethic is positive. Older persons with family and financial responsibilities are likely to espouse a greater commitment to the work ethic than a younger person without such responsibilities.

Only very weak positive relationships were found. It is possible that the work history of the respondents might serve to weaken the expected relation. While detailed work histories were not collected, the fact that the respondents were enrolled or applied for CETA services is evidence of interruption of unsubsidized labor force participation.

The next relationship explored is between the work ethic and earnings before CETA participation or application. Once again the expected relationship of the two variables in question might be expected to differ between a "normal" population, that is, one with an uninterrupted history of labor force participation, and program participants and applicants. For the former a positive relationship between wage and work ethic might be expected. While loss of job might not be expected to effect a permanent reversal of attitudes on the part of higher-waged workers, a temporary disillusionment would result for CETA applicants and participants. It is possible that the negative relationship obtained reflects such feelings.

The possibility that the employment status of program graduates could be related to work-ethic commitment has been alluded to in the preceding discussion. For the three program components, work-ethic means are higher for employed graduates than for unemployed graduates. It is not possible to conclude for the reported scores whether the higher work-ethic support among employed graduates was a cause or a consequence of their employment. Similarly, the data at hand do not permit one to conclude that unemployed graduates are unemployed because of their lack of commitment to the work ethic. In all likelihood the two, work ethic and employment status, are mutually reinforcing.

Another factor expected to be related to commitment to the work ethic is the nature of employment obtained. Do individuals who obtained training-related jobs exhibit greater allegiance to the work ethic than those program graduates who did not get jobs in their field of training? Table 4–2 provides results which answer this question in the affirmative, and also summarizes the work-ethic means for all groups.

It is interesting to compare the results reported for training-related employment with those in the employed category. For all groups the work-ethic means of the training-related placements exceeded those graduates who reported being employed. In addition, program graduates in training-related jobs reported a higher commitment to the work ethic than employed eligible nonenrollees. Even more interesting is the finding that in all cases the work-ethic means of those who did not obtain training-related employment were lower than those of the employed graduates. The scores of the nontraining-related job group, however, do exceed those of the unemployed graduate groups.

The conclusions to be drawn from the preceding discussion must be tentative because firm conclusions regarding the temporal ordering of work-ethic commitment and employment have not been reached. The consistency of the pattern of relationships reported thus far, however, does allow one to construct

Table 4–2
Comparison of Mean Work-Ethic Scores for All Program Participants and Graduates with Those of Eligible Nonenrollees

		OJT		*Classroom Training*		*Work Experience*	
	Eligible Nonenrollees N = 75	*Current OJT N = 43*	*OJT Graduates N = 38*	*Current Classroom Training N = 68*	*Classroom Training Graduates N = 80*	*Current Work Experience N = 86*	*Work Experience Graduates N = 50*
Work-ethic mean	3.24	3.18 (-0.06)	3.15 (-0.09)	3.15 (-0.09)	3.10 (-0.14)	3.13 (-0.11)	2.65 (-0.59)

	Employed Eligible Nonenrollees N = 58	*All Employed OJT Graduates N = 33*	*Training-related OJT Graduates N = 10*	*Non-training related OJT Graduates N = 23*	*All Employed Classroom Training Graduates N = 62*	*Training-related Classroom Training Graduates N = 13*	*Non-training-related Classroom Training Graduates N = 49*	*All Employed Work Experience Graduates N = 27*	*Training-related Work Experience Graduates N = 3*	*Non-training-related Work Experience Graduates N = 24*
Work-ethic mean	3.19	3.18 (-0.01)	3.42 (+0.23)	3.04 (-0.15)	3.13 (-0.06)	3.30 (+0.11)	3.12 (-0.07)	2.75 (-0.44)	3.36 (+0.17)	2.68 (-0.51)

	Unemployed Eligible Nonenrollees N = 17	*Unemployed OJT Graduates N = 10*	*Unemployed Classroom Training Graduates N = 18*	*Unemployed Work Experience Graduates N = 23*
Work-ethic mean	3.34	2.94 (-0.40)	3.00 (-0.34)	2.52 (-0.67)

Notes: Range of values is from 1.0 to 4.0 with 1 0 representing the least commitment and 4.0 representing the highest commitment to the work ethic. Differences between eligible nonenrollees and program participants and graduates are in parentheses.

a plausible interpretation. It appears that mere participation or completion of a CETA program is not sufficient to bring participants' or graduates' work-ethic scores in line with those of eligible nonenrollees. The lower scores of employed graduates compared to eligible nonenrollees indicate that just obtaining a job is not an adequate stimulus to boost work-ethic scores. Only the attainment of a training-related job is enough to boost the graduates' score above that of the eligible nonenrollee group.

The relationships discussed thus far have been bivariate in nature. The inability of this approach to identify spurious relationships and sort out the independent impacts of each variable on the work ethic serves to underscore the tentativeness of any conclusions arrived at thus far. The analysis undertaken in the next section attempts to remedy these deficiencies using a multiple regression technique employing dummy variables.[10]

Analysis

Independent Program Effects

A number of regressions using several combinations of variables were calculated. Table 4-3 presents the results for two such equations. In equation 4.1, in addition to the program dummy variables, three independent variables are included: education, age, and pre-CETA wage.

Two of the three nonprogram coefficients are significant. An increase of $1 in pre-CETA income results in a slight decrease in the work-ethic mean. As previously suggested, it is possible that the loss of a job might prompt a temporary disillusionment on the part of CETA applicants and participants. A slight positive relationship exists between age and work-ethic mean, holding other variables constant.

Inspection of the coefficients for each of the program categories reveals that there is a decrease in the work-ethic mean compared to that for the eligible nonenrollee group for each category. However, only one coefficient, that for work-experience graduates, is statistically significant. The work-ethic mean for a work-experience graduate is 0.57 lower than that for an eligible nonenrollee with the same characteristics.

Equation 4.2 presents the results of a regression with the education variable omitted and an additional dummy variable included, training-related nature of job. The training dummy takes in a value of 1 if post-CETA employment is training-related, 0 if not. As can be readily seen, the coefficient for the training-related variable is positive and statistically significant. Holding other things constant, the work-ethic means for individuals who obtained a training-related job increase by 0.30 over the eligible nonenrollee group. When viewed in conjunction with the coefficients for the program variables, both OJT and classroom-training graduates

Table 4-3
Determinants of the Work-Ethic Mean

Constant	*Education*	*Pre-CETA Wage*	*Age*	*OJT Current*	*OJT Graduate*	*CRT Current*	*CRT Graduate*	*WE Current*	*WE Graduate*	R^2
3.07	0.03 (0.71)	-0.04[a] (4.06)	0.007[a] (6.66)	-0.04 (0.12)	-0.05 (0.21)	-0.05 (0.31)	-0.12 (1.98)	-0.16 (3.00)	-0.57[a] (29.55)	0.10 (5.27)

Equation 4.1
Dependent variable = work-ethic mean
Coefficients (F scores in parentheses)

Constant	*Pre-CETA Wage*	*Age*	*OJT Current*	*OJT Graduate*	*CRT Current*	*CRT Graduate*	*WE Current*	*WE Graduate*	*Training Related Job*	R^2
3.18	-0.04[a] (4.59)	0.005[a] (5.17)	-0.07 (0.41)	-0.12 (1.18)	-0.09 (0.88)	-0.18 (3.76)	-0.19[a] (4.58)	-0.59 (33.27)	0.30[a] (12.75)	.13 (6.76)

Equation 4.2
Dependent variable = work-ethic mean
Coefficients (F scores in parentheses)

[a]Significant at the .001 level.

move from a deficit with respect to eligible nonenrollees to a position of higher scores, that is to say, OJT and classroom-training graduates who obtained a training-related placement will have, holding other variables constant, a stronger commitment to the work ethic than the eligible nonenrollee group.

Results

These results indicate that program participation exerts a negative impact on work-ethic scores for participants and graduates as compared to those of eligible nonenrollees. This finding clearly does not conform to the expectations of either the culture-of-poverty or situational philosophies outlined earlier. The ordering of program participants by category corresponds to that predicted by the situational school: work-ethic scores are highest for OJT participants, followed by classroom-training participants, and then work-experience participants. The superiority of the eligible nonenrollees, however, does not follow from the tenets of the situationalists. While all the groups in this study scored lower on the work ethic than the general-population groups reported by Goodwin, the different modes of data collection and ten-year time lag may explain the differing results.

The expectations derived from the culture-of-poverty school receive no empirical support in this study. Rather than scoring highest, work-experience participants and graduates posted the lowest scores. The conclusions one reaches about title I programs depend upon the assumptions made about intake policies. The best case for the programs is made as follows: if one assumes the participants were less motivated than eligible nonenrollees at the time of program admission, the inability of participants to match scores with eligible nonenrollees could be interpreted as a regrettable but not particularly pernicious result. The program could be regarded as doing a creditable job on a difficult task. This interpretation is undermined, to some degree, by the even lower scores posted by program graduates. It is possible that expectations raised during program participation were lowered by realities faced upon program completion conforming to the transitory model of program benefits.

The worst case for the program is as follows: if creaming is assumed, the lower scores reported for participants indicate a decline since program admission. The even lower scores of graduates than participants paint an unflattering picture of the benefits to be derived from CETA participation with respect to the inculcation or enhancement of favorable attitudes toward work. The opinion has been voiced by some that make-work projects give the participants a distorted image of work. Gurin, in the same vein, asserts that this "danger would be particularly an issue in programs that focus on socialization issues to the neglect of skill and competence training. To encourage the values and beliefs of an internally oriented protestant ethic without training in some of the skills that might help one implement these values is potentially psychologically destructive."[11] To the extent that work experience conforms to this description, the lower work-ethic scores for that group of participants and graduates are put into perspective.

The worst-case scenario is rendered less onerous by the fact that the work-ethic means posted by graduates who obtained training-related placements exceeded those of nonenrollees. Providing marketable skills appears to be one of the most important factors in bolstering commitment to the work ethic. Mere program participation and completion are insufficient to accomplish this task. Moreover, participation in a program that provides an individual only with an introduction to the world of work could have detrimental consequences for belief in the work ethic.

Despite the fact that the rank ordering of groups on the work ethic did not fully conform to the expectations of either school, that predicted by the situational approach was more closely approximated than that of the culture of poverty. This result leads one to conclude that it is faulty implementation, and not faulty theory, that is the problem. The fact that the work-ethic scores of program graduates who have been placed in training-related jobs exceed those of employed eligible nonenrollees underscores this conclusion. The program is successful in enhancing feelings toward work so long as the marketable skills are truly marketable.

The work-experience component presents a problem in disentangling the effects of theory and program implementation. Viewed apart from the other programs, the attribution of blame to the culture-of-poverty theory or program implementation is uncertain. However, in juxtaposition to the other programs, its lowest ranking suggests that the assumptions underlying the program are suspect. It is likely that a perfectly implemented program would fail with regard to enhancement of the work ethic as no marketable skills are obtained by the participant.

Political Attitudes

Political efficacy is one of the more overtly political issues examined in this analysis. The expectation is that CETA participation would result in more efficacious attitudes on the part of current participants and graduates when compared to the eligible nonenrollees. This expectation finds theoretical support in the classical democratic and human-capital theories discussed earlier. Attempts to move individuals into the economic mainstream are held to be accompanied by a parallel movement into the political mainstream.

The Survey Research Center's four efficacy items were used to construct an efficacy index. Inspection of the means of each program component group reveals that all program participants and graduates, with the exception of work-experience graduates, feel more efficacious than eligible nonenrollees. Program graduates are slightly less efficacious than current participants for OJT and work experience; the scores are the same for classroom-training participants and graduates. When the means of employed graduates are compared with all graduates, slightly more efficacious feelings are in evidence. Only employed work-experience graduates espouse less efficacious feelings than employed eligible nonenrollees.

It has been suggested that political efficacy is not a unidimensional concept.[12] Two dimensions have been suggested: one stressing an individual's subjective competence and one illustrating his belief that government is responsive to him. As the former is more under control of the individual, participation in the program might be expected to have a beneficial impact on it. The government-responsiveness dimension, on the other hand, is subject to a myriad of forces and less likely to be altered by program participation. As a result, only the subjective-competence dimension is considered here. The ranking of the program groups remains the same as for the four-item efficacy index. Only work-experience graduates score lower on the subjective-competence dimension than eligible nonenrollees.

The results of a multiple regression analysis reveal that only one program category, current OJT participants, had a beneficial independent impact on either the four-item mean or the subjective-competence mean. The only other variable, apart from education, found to have a beneficial impact was the training-related job variable. A slight increase in efficacy was in evidence for those who obtained training-related jobs.

CETA and the New Federalism

CETA is an example of the New Federalism programs which has as one of its underlying premises that local officials know local needs. The analysis in this section turns that adage around and taps the CETA applicant and participants' perceptions of who runs CETA, what political party gets credit for CETA, and the impact, if any, of these perceptions on their voting habits.

The results of the analysis indicate that considerable confusion exists among participants concerning who operates the program. This confusion is significant to the extent that it indicates a possible flaw in the New-Federalism argument. Local officials may know local needs, but the local population is not clear who is responsible and who should be held accountable. A one-way communication system seems inadequate to the task of ensuring accountability.

No political party emerges as a clear victor. A portion of the respondent's confusion could be due to the fact that the Republicans occupied the White House and the Democrats controlled Congress at the time when the legislation was enacted. However, the size of the "other" and "do not know" categories indicate that no political party is seen as being responsible for CETA. As expected, a greater percentage of those not enrolled in the program fell into the "do not know" category.

As beneficiaries of a program operated at the local level, one might expect a higher percentage of participants than eligible nonenrollees to cast a ballot in the most recent local election. However, the inability of respondents to identify which level of government operates the program casts considerable doubt on this expectation. The results clearly indicate that program participation does not result in greater electoral participation. In fact, all groups report lower percentages voting in the most recent election than usually vote.

A more explicit examination of the impact of CETA participation or knowledge about CETA on voting is to compare the incidence of voting between groups who think locals run the program and those who do not. One might expect to find a higher percentage of those who believe responsibility lies with the local government casting ballots. The results obtained illustrate that this is not the case. The same analyses were performed for each category with the same results obtained.

Conclusions

Sweeping generalizations about the effects of program participation on the noneconomic variables under consideration, while tempting, are not supported by the empirical evidence presented in the analysis. It is appropriate, however, to indicate that the program's immediate effects are more pronounced on work-related attitudes than on the more overtly political attitudes. This is not surprising when one considers the nature and emphasis of the program. The impacts on political efficacy and participation might not surface until a later date.

While the expectation that the program would affect work-related attitudes was confirmed, the nature of the impact was not as predicted. Instead of heightening commitment to the work ethic, program participation appeared to have a detrimental effect. In all cases program participants reported a lower commitment than eligible nonenrollees. Program graduates posted even lower scores. The implications of this result are profound. Individuals who have expressed an interest in CETA, who meet the eligibility criteria, but who are not admitted into the program espouse more positive feelings toward work than similar individuals who are admitted to the program. The elements of the program that account for these adverse feelings are not clear. The nature and length of the training programs could be responsible. The stigma attached to participation in a government-funded program implying that one could not successfully compete in the mainstream labor force could be partly responsible. Unfulfilled expectations raised by the prospect of program participation could also be a factor. While the reasons why the work ethic is lower for CETA participants and graduates than for eligible nonenrollees remain ambiguous, the route to instilling or reinforcing a greater commitment is to ensure that the CETA graduate gets a job. Mere employment, however, is not sufficient to reverse negative attitudes. Only graduates who were able to get training-related employment reported a greater commitment to the work ethic than eligible nonenrollees.

The success of the training-related placements prompts one to conclude that the situational and not the culture-of-poverty theory is more appropriate in designing manpower training programs. While the expectations of the situational school were not totally supported, they were more closely approximated than those associated with the culture-of-poverty school. The dismal showing of work-experience participants and graduates compared to the other program components and eligible nonenrollees indicates that a program designed only to alter work habits has a low probability of success. Individuals need to learn a marketable skill and obtain unsubsidized employment using that skill. The traditional emphasis of work-experience programs precludes such an occurrence.

Due to the more circuitous link between program participation and the political effects explored in this analysis, the absence of strong relationships is not surprising. A longer time frame might be necessary for the impacts, if any, to manifest themselves. While the results obtained for efficacy and participation were not surprising, the implications of two of the other political variables examined are noteworthy. The adage associated with the New Federalism—local officials know local needs—was found to be questionable. While the study did not explore the depth of knowledge on the part of local officials concerning the needs of their communities, it did question respondents about who operated CETA programs. A significant percentage did not know that local officials are responsible for the day-to-day operations of the program. At first glance this lack of knowledge does not seem important. However, if one assumes a two-way communications network is necessary for local officials to know local needs, this result indicates that, at best, only one channel is functioning. It is difficult to see how local officials can be held accountable to their constituents for their actions if recipients of CETA services do not know who operates the program. An important construct of the New Federalism seems weak in the case of CETA.

The noneconomic impacts included in this analysis, and the rather surprising nature of a number of those impacts, underscore the importance of evaluating manpower training programs in multigoal terms. The inclusion of the noneconomic goals considered here has resulted in findings that could have significant impact on the design of subsequent manpower training programs, and the political system as well. If work-experience programs were emphasized at the exclusion of skill-training programs, the results could be a group of individuals who are alienated from the work world and the political realm. It appears not to be sufficient for the bolstering of work-related attitudes to participate in a program and find a job. Individuals who learn a skill and employ that skill in an unsubsidized job upon program completion appear to achieve the greatest noneconomic benefits. Given the potential externalities associated with the noneconomic impacts, policymakers cannot afford to ignore them in either the formulation or evaluation stages of policy.

Notes

1. See Lester Thurow, *Investment in Human Capital* (Belmont, Calif.: Wadsworth Publishing Company, 1970).
2. Ibid., p. 105.
3. Charles Perry et al., *The Impact of Government Manpower Programs: In General, and on Minorities and Women* (Philadelphia: Industrial Research Unit, the Wharton School, 1975).
4. *Federal Register,* vol. 42, no. 201, October 18, 1977.
5. A prime sponsorship can be a state, a unit of local government which has a population of 100,000 or more, or a consortium consisting of local governments which are located in measurable proximity to each other with one meeting the 100,000 population requirement. Funds are allocated to the prime sponsorships on the basis of a formula which takes into account a prime sponsorship's unemployment rate, number of poverty families, and the previous year's manpower allocation.
6. The pretest was mailed on September 23, 1977. The first wave of questionnaires was mailed between September 30, 1977, and October 5, 1977. The second wave of questionnaires was mailed between October 24, 1977, and November 16, 1977. The response rate for the first wave was 73.2 percent and for the second, 26.1 percent.
7. Detailed characteristics of the sample as well as a copy of the questionnaire are available from the author upon request.
8. *Federal Register,* vol. 42, no. 201, October 18, 1977.
9. See Leonard Goodwin, *Do the Poor Want to Work? A Social-Psychological Study of Work Orientations* (Washington D.C.: Brookings Institution, 1972).
10. Dummy variables are created by treating each category of a nominal variable as a separate variable and assigning arbitrary scores for all cases depending upon their presence or absence in each of the categories. Program category, for purposes of this analysis, was used to create the dummy variable. As seven categories of programs existed, six categories of the dummy variable were created. The excluded category, in this case eligible nonenrollees, serves as a basis of comparison for the remaining six categories.
11. Gerald Gurin, *A National Attitude Study of Trainees in MDTA Institutional Programs* (Ann Arbor: Institute for Social Research, 1970).
12. See Herbert Asher, "The Reliability of the Political Efficacy Items," *Political Methodology* 1 (Spring 1974): 45-72. See also George Balch, "Multiple Indicators in Survey Research: The Concept 'Sense of Political Efficacy'," *Political Methodology* 1 (Spring 1974): 1-43.

5 Preparation for Apprenticeship through CETA

Kenneth W. Tolo

Introduction

From colonial times, apprenticeship traditionally has been a primary employment initiative of the private sector in the United States. For the individual, apprenticeship eases the transition from school to adult employment responsibilities; it is a method of long-term skills training combining instruction and practical experience. For employers, apprenticeship provides future skilled personnel able to adapt to production and technological changes. For the economy, apprenticeship is an important means of providing more equitable opportunities in employment and training.[1]

Currently, approximately 450 occupations, ranging from carpenter to cosmetologist, have been recognized as "apprenticeable" and offer in-depth, structured, comprehensive training in the skills of a specific trade. Apprentices are exposed first to simple trade skills and tasks, progressing systematically through more difficult assignments until they have mastered all the functions associated with that trade. All programs integrate practical on-the-job training (OJT) with classroom instruction covering trade theory. They are one to five years in duration, though most last at least three years. Programs pay wages to apprentices and are privately funded. Program sponsorship is of three types: by a single employer, by a group of employers, or by a cooperative labor-management joint apprenticeship and training committee (JATC). An apprentice is recognized as a journeyman upon completion of the program, thereby qualifying for the top wages of the trade paid in that labor market.

The apprenticeship system traditionally has been both white male-oriented and independent, with its accepted, consistent structure of apprenticeship programs often standing in contrast to the changing social, political, and industrial environment in which it now exists. Federal and state apprenticeship agencies are increasingly active, albeit often in an uncoordinated manner, in program registration, servicing, development, promotion, and compliance.[2] Federal regulations on apprenticeship program standards (for example, 29 Code of Federal Regulations (CFR) Part 29) and affirmative action (29 CFR Part 30) impose substantial administrative burdens on apprenticeship sponsors. Recent court decisions still leave unresolved, at least to many employers, the legality of reverse-discrimination approaches to achieve more equal employment opportunities for all workers.[3] The expansion of apprenticeship into new occupations requires nontraditional, innovative program approaches. Yet other challenges are imposed by the educational system, which primarily has focused on college preparation and often has failed to provide graduates with the training and desire necessary to enter apprenticeship programs in the skilled trades, and by male-female role differentiation in early life and in career counseling.[4]

Preapprenticeship programs, or programs which prepare persons for apprenticeship, provide one effective means to meet the needs of both the apprenticeship community and the targeted population. When included in affirmative action plans, for instance, preapprenticeship is viewed by the federal government as a "good faith effort" on the part of the program sponsor to comply with the goals and timetables of 29 CFR Part 30. These programs have long been used in specific crafts to instruct apprenticeship candidates in safe working procedures and to provide them with prejob skills training in order to increase the first-day productivity of the apprentice. When combined with more general instruction on the nature of apprenticeship and the skilled trades, this training can enable greater numbers of women and minority males to compete more effectively for apprenticeship positions.

But no single mix of preapprenticeship program services can be universally applied to all trades. The design of preapprenticeship programs involves a careful matching of the characteristics, aptitudes, and abilities of targeted groups of participants (that is, potential apprentices) with the requirements set by employers and JATC officials. This individualized program development for workers who often lack the skills or knowledge to be productive on the job is costly for the private sector, however, and expansion of these programs has not been rapid. Moreover, general uncertainty about what elements constitute a preapprenticeship program create even further problems in expanding the use of these programs to assist the greater entry of women and minority males into apprenticeship.

The Comprehensive Employment and Training Act (CETA) of 1973 (as amended) offers a largely untapped resource of programs and funds for the expanded development of preapprenticeship programs. Following a review of CETA as a potential resource and of factors inhibiting preapprenticeship/CETA coordination, this chapter offers policy and program recommendations for consideration by the Federal Committee on Apprenticeship (FCA), the U.S. Department of Labor (DoL), the apprenticeship community, and other individuals or groups interested in developing preapprenticeship programs. It focuses solely on CETA-funded programs in order to understand more clearly the diverse ways in which CETA funds are used and how greater integration with preapprenticeship might be achieved. Findings and recommendations, discussed in greater detail in a final report of the 1978–1979 LBJ School of Public Affairs Apprenticeship Project, are based upon an extensive literature review and on-site program interviews in nine U.S. urban areas and the Navajo Indian Reservation (Arizona) with apprenticeship system, CETA system, government, and community agency personnel (see table 5-1 for programs visited).[5]

Table 5-1
Characteristics of CETA-Funded Preapprenticeship Programs: 1978-1979 LBJ School Apprenticeship Project Survey

	Program Components										*Relation to Structured Apprenticeship Programs*						*Target Groups*					*Industry Focus*	
Program Type and Name	*Recruitment*	*Assessment*	*Orientation*	*Counseling*	*Educational Services*	*Skills Training*	*Tool Familiarization*	*OJT*	*Follow-up Services*	*Wage or Subsidy*	*Operated by Apprenticeship Organization*	*Apprenticeship Credit Given*	*Guaranteed Job or Receive Preference*	*Apprenticeship Organization Participated in Design*	*Refers Clients to Apprenticeship*	*Agreement with Apprenticeship Organization*	*Women*	*Black*	*Hispanic*	*Youth*	*Other*	*Building & Construction Trades*	*Other*
Apprenticeship Outreach																							
Recruitment and Training Program, Inc. (Boston; in operation 10 years)	x		x	x	x				x						x		x	x	x	x		x	
Labor Education Advancement Program (LEAP), National Urban League																							May expand into construction trades
Los Angeles LEAP	x		x	x	x				x						x		x	x	x			x	
Chicago LEAP (in operation 13 years)	x		x	x	x				x						x	x	x	x	x	x		x	
United Auto Workers Outreach Program (Detroit; in operation 11 years)			x	x	x				x	x	x						x	x	x		x		Automobile manufacturing
Mexican-American Opportunities Foundation, Youth Apprenticeship Program (Los Angeles; in operation 6 years)		x	x	x	x		x								x				x	x	x	x	

Table 5–1 *Continued*

Program Type and Name	*Program Components*										*Relation to Structured Apprenticeship Programs*						*Target Groups*					*Industry Focus*	
	Recruitment	*Assessment*	*Orientation*	*Counseling*	*Educational Services*	*Skills Training*	*Tool Familiarization*	*OJT*	*Follow-up Services*	*Wage or Subsidy*	*Operated by Apprenticeship Organization*	*Apprenticeship Credit Given*	*Guaranteed Job or Receive Preference*	*Apprenticeship Organization Participated in Design*	*Refers Clients to Apprenticeship*	*Agreement with Apprenticeship Organization*	*Women*	*Black*	*Hispanic*	*Youth*	*Other*	*Building & Construction Trades*	*Other*
Outreach with Skills Development																							
Latin American Task Force (Chicago; in operation 11 years)	x	x	x		x	x			x						x	x			x			x	
Preapprenticeship Training Program (Dallas; in operation 1 year)	x		x		x	x		x		x					x		x	x	x			x	
Women in Construction (Boston; in operation 2 years)	x	x	x	x	x		x	x	x	x			x	x	x		x					x	
Emergency Home Repair (Portland, Ore.; in operation 12 years)	x	x				x		x		x				x	x					x		x	

Navajo Construction Industry Manpower Program (Window Rock, Ariz.; in operation 3 years)	x	x	x		x		x	x	x	x			x	x				x	Navajo Indians	x
Craft Readiness Training King County Carpenters Joint Apprenticeship and Training Trust Preapprenticeship Program (Seattle; in operation 13 years)				x	x	x	x		x	x		x	x		x				CETA-eligibles	x
Carpenters Joint Apprenticeship and Training Fund for Southern California (Los Angeles; in operation 5 years)		x	x	x	x	x		x	x	x			x		x				CETA-eligibles	x
Los Angeles County Building and Construction Trades Council Vestibule Project 408 (in operation 1 year)		x		x	x	x			x	x			x	x	x	x	x			x
Home and Apartment Builders Association of Metropolitan Dallas (in operation 1 year)				x	x	x			x	x	x		x	x					CETA-eligibles	x

Table 5-1 *Continued*

Program Type and Name	*Program Components:* *Recruitment*	*Assessment*	*Orientation*	*Counseling*	*Educational Services*	*Skills Training*	*Tool Familiarization*	*OJT*	*Follow-up Services*	*Wage or Subsidy*	*Relation to Structured Apprenticeship Programs:* *Operated by Apprenticeship Organization*	*Apprenticeship Credit Given*	*Guaranteed Job or Receive Preference*	*Apprenticeship Organization Participated in Design*	*Refers Clients to Apprenticeship*	*Agreement with Apprenticeship Organization*	*Target Groups:* *Women*	*Black*	*Hispanic*	*Youth*	*Other*	*Industry Focus:* *Building & Construction Trades*	*Other*
International Masonry Apprenticeship Trust, Denver Pre-Job Subcontract (in operation 2 years)			x		x	x	x	x	x	x	x	x		x	x		x	x	x		economically disadvantaged	x	
Operating Engineers Local #98, OP-EN PARC (Westfield, Mass.; in operation 2 years)						x	x	x		x			x				x	x	x	x	CETA-eligibles	x	
United Auto Workers On-the-Job Training (Los Angeles)			x					x	x		x			x	x		x	x	x				Automobile manufacturing

Automotive Services Council (Los Angeles; in operation 2 years)		x	x					x		x	x	x			x				Automobile servicing
Colorado Construction Employment and Training Association (Denver; in operation 1 year)	x	x	x					x		x	x	x		x	x		x	CETA-eligibles	
L.H. Bates Vocational-Technical Institute of Tacoma (in operation 31 years)			x		x	x				x			x	x	x			CETA-eligibles	23 apprenticeable trade areas
Hampden District Regional Skills Center Tool and Die Preapprenticeship Program (Springfield, Mass.; in operation 1 year)		x	x	x	x	x	x			x		x	x	x	x	x		CETA-eligibles	Tool & die
National Tool and Die Preemployment Program, Akron Skills Center (in operation 9 years)		x				x			x	x	x			x	x			CETA eligibles	Tool & die

Table 5-1 *Continued*

Program Type and Name	*Program Components*										*Relation to Structured Apprenticeship Programs*						*Target Groups*					*Industry Focus*	
	Recruitment	*Assessment*	*Orientation*	*Counseling*	*Educational Services*	*Skills Training*	*Tool Familiarization*	*OJT*	*Follow-up Services*	*Wage or Subsidy*	*Operated by Apprenticeship Organization*	*Apprenticeship Credit Given*	*Guaranteed Job or Receive Preference*	*Apprenticeship Organization Participated in Design*	*Refers Clients to Apprenticeship*	*Agreement with Apprenticeship Organization*	*Women*	*Black*	*Hispanic*	*Youth*	*Other*	*Building & Construction Trades*	*Other*
Nonpreapprenticeship Programs																							
Boston Marine Industrial Park Skills Training Center (Boston; in operation 3 years)		x	x	x	x	x	x			x											CETA-eligibles		Ship repair & maintenance
Apprenticeship Information Center (Portland, Ore.; in operation 16 years)															x								

CETA as a Resource

CETA is the nation's primary public program for moving unemployed, underemployed, and low-income individuals into productive jobs. Created in 1973 to provide coordinated employment and training services to eligible individuals, CETA simplified the administrative structure of federal employment and training programs by consolidating approximately 10,000 categorical grants-in-aid previously funded under the Manpower Development and Training Act, the Economic Opportunity Act, and the Emergency Employment Act.

Most CETA-authorized activities are administered at the local level by approximately 450 prime sponsors. Prime sponsors–states, units of local government with populations over 100,000, consortia of local governments, the balance-of-state area not in any other prime sponsor–are operating government agencies and thus accountable to local political imperatives. Each prime sponsor must have a comprehensive employment and training plan and a broad-based local planning council to guide its employment and training activity. These sponsors are authorized to conduct or subcontract for programs of classroom training, OJT, temporary work experience, job search assistance, supportive services, transitional public service employment, and other training programs. Other CETA programs are administered by governors (for example, vocational education funds, discretionary funds for model employment and training programs) and the secretary of labor (national programs for specific target groups).

Prior to reauthorization in 1978, CETA programs were characterized by such problems as changing (and often conflicting) federal objectives, widespread news reports of alleged fraud and misuse of funds by prime sponsors, lack of private-sector involvement in CETA planning and programs, irregular reporting of positive placements, and substitution by local governments of federally funded CETA participants for normal operating personnel.[6] To correct these problems, Congress limited CETA prime-sponsor flexibility by tightening eligibility requirements, reducing funding for and participation in public service employment, virtually eliminating OJT in the public sector, and mandating the involvement of the private sector through newly established private industry councils (PICs) in each prime-sponsor area. Although the negative image of the CETA system held by apprenticeship officials has not been eliminated, the CETA amendments of 1978 do encourage the private sector to become more actively involved in the development of unsubsidized employment opportunities for CETA-eligible individuals.

Preapprenticeship programs, with their emphasis on enabling greater numbers of women and male minorities to compete for skilled, unsubsidized jobs, offer significant potential for integration with this refocused CETA. Although not every female and minority male is eligible for CETA services (even though some employers wish they were) and not every CETA participant is or should be a female or a minority male, there is sufficient overlap of the participants in the two systems that CETA can be used by the apprenticeship system as a resource for preapprenticeship. Apprenticeship programs gain applicants familiar with the trades and less likely to drop out, while CETA-funded preapprenticeship programs ease the paucity of women and minority male applicants for apprenticeship programs in some trades.

Specific ways in which CETA programs and funds can be linked with apprenticeship programs include (with relevant CETA title in parenthesis):

1. Apprenticeship agency staff and JATC representatives may serve on prime-sponsor and statewide-planning councils, thereby directing local and state CETA program activities (title I).
2. Employers may use CETA upgrading and retraining programs to prepare current and potential employees to enter apprenticeable occupations (title II).
3. Prime sponsors may "buy into" existing preapprenticeship programs (placing a CETA-eligible participant in prejob training with unsubsidized participants) (title II).
4. Prime sponsors can fund entire classroom training or OJT programs designed to serve the preapprenticeship needs of specific crafts (title II).
5. Governors may establish model training programs for preapprenticeship purposes (title II).
6. Temporary public service employment may be combined with training to prepare individuals for apprenticeship (titles II-D and VI).
7. National OJT and targeted outreach programs may be used by specific trades to prepare targeted groups to enter apprenticeship (title III).
8. Although more an outreach effort than preapprenticeship, summer youth programs may be used to introduce young men and women to the opportunities and fundamentals of the skilled trades (title IV).
9. Job Corps equipment, facilities, and curricula may be used for preapprenticeship programs (title IV).
10. Youth employment demonstration projects may involve prime-sponsor and apprenticeship linkages through preapprenticeship, and apprenticeship officials also may serve on local and state youth councils (title IV).
11. Apprenticeship officials may serve on PICs, thereby helping to focus CETA programs more on private-sector employment (title VII).

Factors Inhibiting Coordination

Although the use of CETA resources for preapprenticeship is to a large extent limited only by the creativity and understanding of CETA officials and members of the apprenticeship community, joint program efforts have not developed rapidly. Moreover, the descriptive literature on specific preapprenticeship programs has provided little insight into the general types and characteristics of

CETA-funded programs or the broader actions which might be taken to improve coordination. An important step in gaining this insight is to understand the factors that currently inhibit coordinated programs.

One important factor has been the absence of a single, widely accepted definition or typology for the term "preapprenticeship." This ambiguity has hampered the systematic establishment of programs to prepare individuals for apprenticeship. Many practitioners have viewed preapprenticeship programs as training for apprenticeship in a specific trade; others have been more concerned that preapprenticeship programs should prepare women and minority males to compete successfully in apprenticeship entry processes; others have viewed preapprenticeship programs as improving occupational safety and as increasing apprentice productivity and retention. Some practitioners interpret the prefix "pre" to mean "coming before," thereby giving an implied promise to individuals enrolling in a preapprenticeship program; but the LBJ School of Public Affairs study found that guaranteed admission to apprenticeship for completers of preapprenticeship programs is an exceptional circumstance.[7]

In 1964 an FCA task force proposed a preapprenticeship typology that included three program groupings.[8] One category (preapprenticeship) encompassed theoretical training, without pay, completed prior to an apprenticeship position and with no guarantee of an apprenticeship. Prejob apprenticeship included training programs for a specific trade, with or without wages, with completers employed as apprentices. The third FCA category (training for apprenticeship) consisted of on-the-job programs that provided an extended apprenticeship probationary period. Although each of these three types of programs exists today, since 1964 there has been a shift in the nature of preapprenticeship to greater emphasis on outreach activities. Hence the FCA categories no longer are sufficiently inclusive and unambiguous.

Responses in the LBJ School of Public Affairs study to the question, "what is preapprenticeship?" confirmed this continuing ambiguity:[9]

> Preapprenticeship refers to a familiarization with a particular trade so that the trainee will be productive on the first day of apprenticeship (regional director, Bureau of Apprenticeship and Training, U.S. Department of Labor).
>
> Preapprenticeship is compensatory education. It is a program that improves skills people are lacking in order to give them a headstart in apprenticeship. Preapprenticeship is training to compensate for limited or no background in the trades (Apprenticeship Program director).
>
> Meaningful skilled training that labor unions conduct in conjunction with an employment and training agency to bring disadvantaged clients "over the hump" into apprenticeship with a union is the basic element of preapprenticeship (State Employment and Training Council official).

> Today there are two types of preapprenticeship programs: those that are connected with a trade and lead directly into an apprenticeship program (and are a success); and those general programs not directed at any one trade and without any guarantee of a job or an apprenticeship slot (Building and Construction Trades Council member).

Systemic differences further inhibit productive CETA/apprenticeship working relationships. Officials in each of the two systems generally are unaware of the resources, operations, and objectives of the other system, yet they do not hesitate to express their attitudes about the other system's program, and these views often have been negative. One basis for these attitudinal barriers has been the historic tension between the Bureau of Apprenticeship and Training (BAT) and other DoL agencies and offices within the Employment and Training Administration (ETA). Many BAT staff, most of whom are closely associated with the apprenticeship tradition and skeptical of government-funded employment programs (especially CETA), speak of the "good old days" when they had the authority to distribute funds and monitor programs. Tension caused by the subsequent shift of responsibilities still exists at the local level, where historically there has been little incentive for contacts (let alone jointly operated programs) between prime-sponsor staff and local, state, or regional BAT (and state apprenticeship agency) personnel.

Apprenticeship officials tend to stigmatize CETA-eligible individuals, viewing them as incapable of successfully completing apprenticeships. CETA training is widely thought to be of poor quality, inadequate in length and depth, and ignorant of the needs of the skilled trades. These officials view CETA programs as producing large numbers of workers trained without regard to labor market demands, who create downward pressures on wages and glut the labor market with semitrained workers at a time when even highly skilled workers are suffering high unemployment.

From the CETA perspective, prime sponsors often are not informed about apprenticeship programs in their jurisdictions. Even when they are aware of these programs, many CETA officials view the apprenticeship system as closed to CETA-eligible women and minority males, either because of the long-standing tradition of male dominance in the skilled trades or because of perceived racist and sexist attitudes in the apprenticeship community. Prime sponsors, who believe their major concern is placing their participants in unsubsidized positions, sometimes feel it is easier to develop totally new training programs to meet this objective than to attempt to build on existing apprenticeship programs.

Structural barriers to the use of CETA resources by the apprenticeship system also exist.[10]

1. Disparate backgrounds of officials in the two systems (former apprentices as compared with recent college graduates) may lead to misunderstanding and difficulty in communication.

2. Different jurisdictional boundaries in local labor markets (JATC compared with prime sponsor) compound the administrative barriers to coordination.
3. The apprenticeship system is not monolithic, that is, each trade has its own entry process, making it difficult for individuals outside the apprenticeship community to understand and work with the wide variety of hiring practices and organizations.

Regulatory barriers to the use of CETA funds for preapprenticeship include income eligibility requirements, payments to participants, and allowable costs. Although CETA income criteria vary somewhat by programs, with titles II, III, and IV containing some exceptions, most CETA-eligible individuals must be unemployed, underemployed, or economically disadvantaged. With respect to participant payments, CETA programs (other than OJT) generally have not funded private-sector training for participants in unsubsidized jobs, for example, a probationary preapprenticeship period. Moreover, not all cost components of preapprenticeship programs are allowable for CETA funding. For example, programs with a multitrade training component (masonry, carpentry, electrical trades, plumbing) find the cost of equipment to be a problem.

While strict adherence to CETA regulations may inhibit the use of CETA funds for preapprenticeship, at other times regulations designed to encourage that use are ignored. Two such examples are the membership of CETA advisory councils and consultation regarding training programs in apprenticeable occupations. With respect to the former, complaints about the dominance of community-based organizations and the lack of labor input on these councils are common among apprenticeship officials. In addition, prime sponsors often are criticized for operating training programs in apprenticeable occupations without informing local BAT or state apprenticeship agency staff.

Further hindering intersystem cooperation have been procedural barriers, which include on the one hand CETA paperwork requirements and monitoring practices and on the other apprenticeship entry requirements and infrequent apprenticeship program openings. Program operators object to the extraordinary amount of paperwork and documentation required by CETA prime sponsors, with their frustration deepening when information provided to CETA officials is not used. In addition, prime-sponsor program monitors sometimes are alternated frequently, inhibiting the establishment of effective relationships between the apprenticeship program and the prime sponsor. (Exceptions among programs surveyed in the LBJ School of Public Affairs study include the Colorado Construction Employment and Training Association (C-CETA) in Denver and a few national programs funded by the DoL Office of National Programs.[11]

Apprenticeship system procedures also hinder the use of CETA resources for preapprenticeship. A maximum age limit still is a barrier in some trades. Moreover, the fact that most apprenticeship programs accept new entrants only once or twice a year has made it difficult for CETA officials to provide applicants.

Without special programs designed for interim work experience or preparation, prime sponsors have been unable to refer CETA-eligible applicants to apprenticeship programs that have not yet begun.

Recommendations for Improving Coordination

The summary of recommendations that follows seeks to clarify the general role and service components of CETA-funded preapprenticeship programs, thereby providing guidance both on the improvement of greater employment equity in existing programs and on the establishment of new programs whose service components are consistent with program intent.

1. If preapprenticeship is to become a meaningful concept, the various types of preapprenticeship programs must be identified and agreement must be reached regarding a standard typology. Because use of the term *preapprenticeship* varies greatly within the apprenticeship community, the utility of the term has diminished. It should not be used to refer to a specific type of program of preparation for apprenticeship; rather it should only be used as a generic term to refer to all programs which prepare individuals directly and specifically to compete for apprenticeship positions.

The diverse types of CETA-funded preapprenticeship programs currently in operation generally can be grouped into three categories by analyzing the services that each program provides: amount and type of craft training provided, emphasis on supportive services, and extent of association with recognized apprenticeship programs (see table 5-1).

Apprenticeship Outreach. The primary function of Apprenticeship Outreach programs is to recruit individuals for apprenticeship and to prepare them to compete more successfully for entry into apprenticeship. Services toward these ends may include provision of information regarding apprenticeship training and application procedures as well as instruction designed to prepare individuals for apprenticeship-entry tests and interviews. Apprenticeship Outreach programs typically are not trade-specific, but provide a broad range of services applicable to many trades. These programs conduct no trade skill training as such. They tend to be oriented toward specific groups of individuals which historically have been underrepresented in apprenticeable crafts. Because of this focus on the clients, Apprenticeship Outreach offices usually are located so that the client group has easy access to program services.

Outreach with Skills Development. Outreach with Skills Development programs provide recruitment and information services similar to Apprenticeship Outreach programs. In addition, these programs include training designed to increase manipulative skills and to provide clients with a threshold proficiency in the skills of a certain trade. Training is usually in the form of hands-on training complemented by craft-related classroom instruction, though programs may include orientation to the work processes of a specific trade.

Craft Readiness Training. The primary emphasis of Craft Readiness Training is upon skill development for a specific craft. Recruitment and orientation to apprenticeship are not major concerns. Training services may take the form of OJT under the supervision of an employer, or may be in the form of hands-on training offered in central locations such as skills centers or vocational schools. Craft-related instruction in a classroom setting may also be a component of Craft Readiness Training programs. Training periods are long enough to allow for intensive skills training, but this training does not serve as a substitute for a structured apprenticeship program. These programs seek to provide trainees with the skills necessary to get a good start when, and if, they become apprentices.

Each of these classes of programs has an explicit focus on apprenticeship, yet each serves special purposes and meets the needs of different target groups and industries. Apprenticeship Outreach programs, initially designed to recruit minority males for apprenticeship positions in the building and construction trades, generally are affiliated with national organizations but based in the community or neighborhood from which their clientele are drawn. These programs tend to have extensive contact networks from which highly motivated participants can benefit, but normally provide neither work experience nor a guaranteed apprenticeship position. On the other hand, Craft Readiness Training programs stress the development of a person's craft skills relevant to a specific trade, with little emphasis on support services. They are designed to prepare participants to perform productively as apprentices. Outreach with Skills Development programs combine the recruitment, orientation, preparation, and counseling services of Apprenticeship Outreach programs with the trade-specific emphasis of Craft Readiness Training programs that goes beyond basic tools familiarity.

Preapprenticeship programs are built of service components, with the absence or presence of these components determining the category into which a program is placed (see table 5-1). Program service components generally include (to a greater or lesser degree) recruitment, intake, and assessment; orientation and counseling; educational services; supportive services; and skills development. Special components for target populations (physical conditioning and "assertiveness training" for women; bilingual instructors) and provisions for apprenticeship credit may be provided.

In each program included in table 5-1, the mix of services provided represents a serious effort to enable certain workers to compete for entrance into apprenticeship who otherwise would be unable to qualify. Only three (of the twenty-four) programs guarantee apprenticeship positions to preapprenticeship program participants, yet the placement of successful program completers in available apprenticeship slots is a common and serious concern.

This proposed preapprenticeship typology is intended to guide members of the apprenticeship community in establishing new programs to prepare individuals, especially women and minority males, to compete more successfully for apprenticeship positions. After an identification of the program category having program objectives and characteristics most similar to those desired of the new program, table 5-1 can suggest components to include in the program structure. Although neither service evaluation nor goal evaluation was possible in the LBJ School of Public Affairs study due to project constraints, this checklist approach can help clarify program goals and provide a starting point for discussions with target-group representatives and with CETA and apprenticeship officials about specific needs and resource availability.[12]

2. There exists no "best" preapprenticeship program design; rather, programs should be designed to meet the needs of applicant (client) groups to which they are adapted. An orientation component could acquaint program staff with cultural, ethnic, and gender-based differences of the target group. Group counseling sessions could bring potential apprentices together with female and minority male apprentices and journeymen to discuss hazing and the level of work during the first few months of apprenticeship. Training in tool identification and trade terminology could be offered on a bilingual basis (for example, for Hispanic clientele).

Programs that work with unemployed and low-income persons frequently need to be concerned with provisions for transportation to the job site. Interim jobs could be provided for unemployed persons encountering a time lag between the completion of a preapprenticeship program and the availability of appropriate apprenticeship openings. In some trades, newly accepted apprentices who previously have been unemployed or underemployed could be given the basic tools or other necessities (special clothing) by preapprenticeship program sponsors.

Programs aimed at assisting and encouraging women to enter apprenticeable crafts need to consider child-care arrangements. Physical training also could be offered to potential women apprentices in need of additional conditioning.

3. Because all types of preapprenticeship programs form a bridge between employers and clients, preapprenticeship program success also requires close relationships with industry. Without direct ties to industry, preapprenticeship programs can get out of step with the labor market, producing more graduates than the market can accommodate or preparing individuals for jobs which do not exist. Apprenticeship Outreach programs, for instance, need to solicit more actively employer and labor estimates of expected job openings on which to base placement goals.

Industry involvement is an important ingredient from the beginning conceptual stages through implementation of the project; it is especially needed in the early discussions about design and the occupational focus of the program.

Agreements need to be reached prior to program implementation about procedures for placement as well as, if appropriate, admission preferences and provisions for advanced credit toward completion of an apprenticeship. As the project gets underway, employers and JATC members need to become acquainted with the activities of the program and with the aspiring applicants. Industry also needs to be represented in staffing the program. Instructors chosen to teach in the classroom and on the job should come from the ranks of qualified journeyman craftworkers.

4. If CETA is to become an effective resource for funding preapprenticeship, actions must be taken by the U.S. Department of Labor and the agencies it funds to encourage improved relationships between the CETA and apprenticeship systems.

In many crafts, industry faces serious challenges in finding qualified candidates for apprenticeship who are motivated, aware of the trade they aspire to enter, and well-prepared to pass the entry requirements and to complete successfully their apprenticeships. Even trades with overall surpluses of applicants generally acknowledge that the numbers of qualified women and minority male candidates is insufficient to meet the goals and timetables outlined in federal regulations. In many local areas, the problem is acute.

The situation is not hopeless, however. By helping to prepare women and minority male candidates for apprenticeship entry and training, preapprenticeship offers an effective strategy for broadening participation in the craft without tampering with job requirements or apprenticeship training standards. Preapprenticeship is not new; on the contrary, it has been used for decades in some trades (bricklaying). Further, public funding for preapprenticeship is available to industry through CETA. To date, CETA is a resource which has not been fully used by apprenticeship sponsors. Due to structural, attitudinal, and regulatory barriers, relatively few preapprenticeship programs use CETA funds. This is particularly true with respect to locally funded projects.

Specifically, staff of BAT should be consulted by DoL CETA staff and local CETA agencies regarding apprenticeship-related program contracts. CETA prime sponsors offering apprenticeship-related training programs should designate a staff liaison officer who is familiar with apprenticeship to work toward better coordination between apprenticeship officials and CETA prime-sponsor staff. CETA prime sponsors should also strive to involve apprenticeship training coordinators and JATC members on their various advisory councils, including PICs. Finally, audience-specific educational campaigns must be developed and delivered to CETA staff and to apprenticeship officials to increase their understanding of the other's sytem.[13]

5. Achieving fuller participation for women and minority males in apprenticeship can be realized by working with schoolteachers and counselors and through greater funding for preapprenticeship.

BAT and state apprenticeship agencies/councils should intensify efforts to educate school counselors and teachers about career opportunities in the skilled trades. Activities which would provide school officials with information about preapprenticeship and apprenticeship could include: (a) extension of the practice of inviting high school counseling staffs to apprenticeship functions; (b) workshops for school counselors to be conducted by BAT and state apprenticeship agencies/councils for the purpose of promoting apprenticeship; and (c) distribution by BAT and state apprenticeship agencies/councils of printed materials and displays on apprenticeship and preapprenticeship suitable for use in high schools.

To expand the placement of female CETA participants in high-paying, unsubsidized jobs, national CETA program staff and local CETA prime sponsors should direct more funds toward preapprenticeship programs for women. The secretary of labor also could consider expanding the use of CETA title III funds available to him to fund preapprenticeship programs targeted toward women and minority males.

6. In conjunction with increasing the availability of funding for preapprenticeship programs, greater effort needs to be devoted to evaluating these programs. Of course, since preapprenticeship programs differ, the evaluations must be tailored to the particular design and goals of each individual program. There appears to be wide diversity in performance of various preapprenticeship programs. Better evaluation could help to identify what works and why. Finally, it should be recognized that preapprenticeship efforts by themselves are no panacea solutions to the problems of equal opportunity.

Notes

1. *Policies for Apprenticeship,* Organization for Economic Cooperation and Development (OECD), Paris, France, 1979, p. 7.

2. *Vol. 2: Coordination of State and Federal Apprenticeship Administration* (A report of the 1978–1979 Apprenticeship Project), Lyndon B. Johnson School of Public Affairs, University of Texas at Austin, 1980.

3. See the June 1979 U.S. Supreme Court decision in United Steelworkers v. Brian V. Weber et al. This decision upheld an alleged reverse-discrimination plan put into effect in 1974 by the Kaiser Aluminum and Chemical Corp. and the United Steelworkers of America (AFL-CIO) to increase the representation of women and minority males in skilled jobs.

4. See Jan Hilton, *Women in Apprenticeship: The Effects of Goals and Timetables in Texas,* Independent Research Project, Lyndon B. Johnson School of Public Affairs, University of Texas at Austin, 1979; Herbert H. Meyer and Mary Dean Lee, *Women in Traditionally Male Jobs: The Experience of Ten Public Utility Companies,* Research and Development Monograph no. 65, U.S. Department of Labor, Washington, D.C., 1978; Roslyn D. Kane, Elizabeth Lee, and

Jill Miller, *Problems of Women in Apprenticeship,* Rj Associates, Arlington, Virginia, 1977; Norma Briggs, *Women in Apprenticeship–Why Not?* Manpower Research Monograph no. 33, Manpower Administration, U.S. Department of Labor, Washington, D.C., 1973; *High School Students View Apprenticeship,* Manpower Administration, U.S. Department of Labor, Washington, D.C., 1971.

5. This work is based on *Vol. 1. Preparation for Apprenticeship through CETA,* a report of a preapprenticeship/CETA study conducted by the 1978-1979 Apprenticeship Project, Lyndon B. Johnson School of Public Affairs, University of Texas at Austin, for the Federal Committee on Apprenticeship and published by the LBJ School in 1979. Members of the LBJ School research team for this study were the author (project director), Professor Robert Glover, and thirteen LBJ School graduate students: Pablo Collins, Paul Hilgers, Alice Kuhn, Annette Lovoi, Anjum Malik, Mark Malnory, Brooks Myers, Mike Nichols, Maria Orozco, Jennifer Pfiester, Don Saylor, Ed Sierra, and Ida Stewart. Project funding was provided by the Lyndon Baines Johnson Foundation, through the LBJ School of Public Affairs.

6. Juan Cameron, "How CETA Came to be a Four-Letter Word," *Fortune,* April 12, 1979.

7. *Vol. 1: Preparation for Apprenticeship.*

8. Meeting Minutes, Task Force Subcommittee on Preapprenticeship, Federal Committee on Apprenticeship, May 18-19, 1964.

9. *Vol. 1: Preparation for Apprenticeship.*

10. For a discussion of entry requirements and procedures of construction industry apprenticeship programs, see F. Ray Marshall, Robert Glover, and William Franklin, *Training and Entry into Union Construction,* Manpower Research and Development Monograph no. 39, U.S. Department of Labor, Washington, D.C., 1975.

11. *Vol. 1: Preparation for Apprenticeship.*

12. Ibid.

13. For recent federal efforts to improve coordination, see *Apprenticeship and CETA Technical Assistance Guide,* Employment and Training Administration, U.S. Department of Labor, Washington, D.C., 1979.

Part II
Equal Employment Opportunity

6 Presidential Leadership, Governmental Reorganization, and Equal Employment Opportunity

Charles M. Lamb

Throughout American history employers have discriminated against minorities and women in hiring and promotion.[1] Although we still fall far short of the ideal of equal employment opportunity (EEO), substantial progress against employment discrimination has nevertheless been achieved in the 1970s.[2] It is therefore crucial that we ask the question: how can discrimination be further reduced by federal employment policies that affect not only the federal civil service but also private, state, and local governmental employment.

The answer in large measure involves government reform initiated through presidential leadership. Under the United States Constitution, the final responsibility for effective civil rights enforcement as well as the enforcement of all other federal laws lies with the president.[3] The president thus has the ultimate duty to provide legal, institutional, and moral leadership to enforce civil rights guaranteed by the Constitution, statutes, and judicial decisions. In the words of Richard Longaker, "[w]here the law assists minority groups, as do the civil rights statutes and many court decisions, [the president] must make clear without cavil or doubt, that it will be enforced."[4] Or in the words of the U.S. Commission on Civil Rights, absent presidential leadership in civil rights enforcement there develops "a steady erosion of the progress toward equal rights, equal justice and equal protection under the Constitution . . ."[5]

While public opinion polls in 1979 showed that President Jimmy Carter's leadership qualities were not widely recognized, he has been seriously attempting to provide effective and forceful leadership to reduce employment discrimination. Soon after his election the president created a task force for reorganizing equal employment enforcement. The task force's reorganization options were presented to Carter in the fall of 1977. After considering the alternatives, which were never officially released to the public, Carter forwarded his equal employment reform proposal to Congress on February 23, 1978. "Reorganization Plan No. 1" was approved by Congress on May 5, 1978–a policy package that the president called "the single most important action to improve civil rights in the past decade."[6]

Through his plan President Carter transferred to the Equal Employment Opportunity Commission (EEOC) enforcement responsibilities previously held by several different federal agencies. In addition to EEOC, the agencies most affected by the reorganization were the Civil Service Commission (now known as the Office of Personnel Management), the Department of Labor, and the Equal Employment Opportunity Coordinating Council (which was abolished). Reorganization Plan No. 1 shifted to EEOC the Civil Service Commission's jurisdiction over cases involving federal job discrimination, the Department of Labor's responsibilities for enforcing the Equal Pay Act and the Age Discrimination in Employment Act, and the Equal Employment Opportunity Coordinating Council's duty to coordinate all federal equal employment activities.[7] However, the Department of Labor continues to oversee federal contract compliance, and the Department of Justice retains its power to sue state and local governments that discriminate in their employment practices.

Three themes are addressed here. The first is a brief examination of the civil rights task force whose members were capably chosen and led by President Carter. Second is a more detailed identification of the president's reorganization policy options and suggestions as to why the Carter Administration made the decision to establish a "Super-EEOC." And third there is a preliminary evaluation of EEOC's recent internal reorganization to deal more efficiently with equal employment complaints and the steps that EEOC has taken toward assuming the prior EEO responsibilities of the Civil Service Commission, the Department of Labor, and the Equal Employment Opportunity Coordinating Council.

President Carter and His Reorganization Task Force

President Carter's dedication to equal employment opportunity was obvious in the initial weeks of his administration. One of the earliest signs came in February 1977, when Carter stated that in the area of employment he was considering consolidating into one agency the "seven major Federal agencies responsible for elimination of sex and race discrimination."[8] Although the president anticipated the consolidation of seven agencies into one, he did not specify which civil rights agencies he had in mind, other than the Civil Service Commission. Presumably the president was suggesting at that early date the consolidation of the EEO responsibilities of the Civil Service Commission and the Department of Labor with EEOC. Most interesting is the fact that Carter's February statement was in response to a very general question concerning future governmental reorganization. That the president chose to speak to reorganization specifically in equal employment policy rather than in other major possibilities such as energy or the Executive Office of the President indicates that Carter had been giving equal employment reorganization particular attention for some time.

Then in late March 1977, Congress authorized President Carter to reorganize the government, with his reorganization plans to go into effect unless within sixty days they were rejected by either house of Congress.[9] Under this law

Carter had three years to submit reorganization plans. The president could withdraw a plan within the sixty-day period, or he could amend it within thirty days after its submission. Each reorganization plan was to address a separate subject matter, such as equal employment opportunity, and could not be amended by Congress.

Soon after receiving the power to reorganize the government, President Carter's former Office of Management and Budget (OMB) director, Bert Lance, announced that Howard Glickstein, law professor at Howard University and a former staff director of the U.S. Commission on Civil Rights, would head the new civil rights task force within the reorganization group at the OMB. According to Lance, Glickstein's civil rights group was to examine equal employment enforcement first and to focus on "fragmentation of responsibilities, lack of accountability, duplication, inconsistent standards and excessive paperwork."[10] Reports in June were that Glickstein's task force would have its employment reorganization options and recommendations to Carter in October 1977.[11]

For making recommendations to Carter, Glickstein and his associates on the task force received inputs from various government agencies, civil rights experts, and interest groups on the question of equal employment policy. Glickstein commented that the president "made a commitment in the campaign to eliminate fragmentation and overlap" in equal job opportunity and that Carter was "anxious to do something about this."[12] He gave a clue as to the future reorganization by suggesting that the creation of one civil rights agency in the area of employment enforcement was an option with "some merit" and that the president's reorganization power was probably a more preferable approach than new legislation for improving enforcement.[13]

The presidential leadership exercised by Carter won virtually unprecedented praise from the Commission on Civil Rights, while Presidents Nixon and Ford had been criticized for their equal employment policies.[14] In fact, the praise from the commission was so strong and supportive that it deserves quotation here. The commission concluded:

> In recent months, increased activity and concern for equal employment opportunity at all levels of the Federal Government have created a basis on which to rest hopes for a greatly improved Federal enforcement effort in the future. First, in February 1977, within three weeks of taking office, President Carter spoke of his intention to give priority to improving the Government's civil rights enforcement effort. In particular, the President noted that there are a number of agencies responsible for implementing equal employment opportunity requirements and stated that it was his goal to move toward a consolidation of these functions. President Carter's subsequent creation of a civil rights reorganization task force within the Office of Management and Budget demonstrates his strong commitment to end employment discrimination.
>
> Second, and of equal importance, is the fact that the officials the President and his Cabinet members have appointed to head civil rights

> programs have, without exception, taken their tasks seriously. Indeed, the summer of 1977 may go on record as the period of greatest activity by civil rights agencies and offices since the Government established mechanisms to combat employment discrimination. The recent appointees have guided their agencies toward renewed efforts at interagency coordination, as is evidenced by the recent activity to develop a uniform set of employee selection guidelines for the Federal Government. Moreover, under the aegis of new leadership, agencies have demonstrated a renewed will to enforce the law firmly. . . .
>
> Third, and perhaps most significant, is the fact that under new leadership several agencies have openly engaged in a critical self-examination of their programs to protect against employment discrimination. The Civil Service Commission, the Office of Federal Contract Compliance Programs, and the Equal Employment Opportunity Commission have all proposed major changes in their operations as a result of self-audits, and many of the provisions in the proposals appear to offer promise for significant improvement.[15]

This report was released by the Civil Rights Commission in December 1977. However, it did not address the optional governmental reforms that should be implemented in the field of EEO. It is to this subject that we next turn our attention.

Presidential Reorganization Policy Options

We now examine the policy options that President Carter had available to him in reorganizing equal employment opportunity. Those of us outside Carter's civil rights reorganization task force have little knowledge of how the administration's strategy actually evolved. We may suggest, though, how such an endeavor should have been systematically undertaken, alternatives that should have been weighed, and advantages or disadvantages of each.

These policy options involve gradations of change and complexity, and thus may be viewed as falling along a continuum of minimum to maximum change involving past EEO structures.[16] The three major groupings of policy options are (1) improving equal employment programs through vigorous new efforts to increase coordination through the traditional enforcement system, with minimal structural change; (2) improving enforcement through the reorganization plan ultimately presented to Congress for the consolidation of EEO structures and accompanying shifts in enforcement responsibilities; and (3) reorganization entailing new legislation, beyond the introduction of a reorganization plan, to abolish some structures and to replace them with new ones possessing greater enforcement powers.

Coordination

Achieving effective coordination through past enforcement structures would have been difficult for President Carter. As the Advisory Commission on Intergovernmental Relations noted, coordination involves a "conservative strategy, and, as such, is often an alternative to the development of more far-reaching social and institutional reforms."[17] Indeed, meaningful reform in EEO could have proven impossible because of the basic problem in the coordination approach: the coordinator normally has no authority to impose uniform standards on programs of other agencies. Nevertheless, the administration had coordinative policy options available to it. The options might have included, for example, coordination by the Department of Justice, the OMB, a Council on Equal Employment Opportunity within the Executive Office of the President, through the Federal Regional Councils, or by the U.S. Commission on Civil Rights.

Option one would have placed the coordination responsibility on the Department of Justice, which is already involved in employment discrimination litigation. In fact, a coordination function was located in the Justice Department during the Johnson Administration. That experience resulted in some coordination in training activities and enforcement procedures, but weaknesses in the approach were also readily apparent.[18] The Department of Justice was and remains oriented more toward litigation than problems of administration and enforcement. Beyond that, agencies were naturally slower to act upon the coordination initiatives of the Department of Justice as opposed to directives from their own officials. In retrospect there is little reason to believe that these conditions have substantially changed, and thus the Carter Administration could quite easily eliminate this option.

Option two would be to delegate the coordination function to the OMB. This might have been advantageous since OMB's jurisdiction cuts across all agency lines and since it exercises influence over all federal programs through the budgetary process. However, OMB's budget examiners have been uninformed on equal employment problems in the past, and they have failed to point out potential civil rights implications when working with agencies on their budgets. In 1977 the U.S. Commission on Civil Rights thus concluded that OMB's budgetary process "was not adequately utilized for the review of Federal civil rights enforcement."[19] In view of this weakness and Carter's stated goal of reducing the size of the Executive Office of the President, OMB was an unlikely location for an equal employment coordination unit.

Had he fallen short of his goal, also within the Executive Office Carter could have established a structure similar to the President's Council on Equal Opportunity created in 1965 by Lyndon Johnson.[20] The Council on Equal Opportunity had the advantage of being composed of agency leaders who could push from within their own organizations to coordinate enforcement, promote improved information flow and contacts among civil rights officials, and initiate similar compliance forms and joint training activities. However, this alternative did not provide timely coordination in the final analysis. Delays occurred as high-level officials, largely unfamiliar with the specifics of their agency enforcement programs, had to be regularly briefed, then had to work with other officials who were assigned overlapping responsibilities, and then had to inform agency enforcement personnel of any agreements or partial solutions reached by the council.[21] This type of delay would similarly result if civil rights coordination were made a job of the more recently established Domestic Council. The Domestic Council is composed of the president and vice president, the heads of most major federal agencies, and prominent officials in the Executive Office of the President. Its duties involve formulating and coordinating recommendations concerning the entire range of domestic priorities. Civil rights coordination would be only one of a multitude of important policy issues to receive attention, thus probably delaying meaningful reform in EEO.

For another example, the Federal Regional Councils (FRCs) could play a greater coordination role in equal employment enforcement.[22] The FRCs are composed of eleven of the most powerful federal agencies in each standard federal region, and all member agencies have various enforcement responsibilities. While this suggestion has potential, coordination would come from the regional rather than the national level, and the councils have no formal authority to force member agencies to cooperate in their enforcement activities. Nevertheless, President Carter announced in September 1977 that the FRCs will continue general civil rights coordination functions in the future.[23]

As a final illustration, the Commission on Civil Rights has been suggested as a coordinating agency.[24] The commission does possess a professional staff well versed in equal employment enforcement, it does conduct enforcement studies and monitor civil rights compliance, and it does serve as a clearinghouse for civil rights information. On the other hand, the Civil Rights Commission is only a temporary agency with no enforcement power. New legislation would probably be required to enable it to oversee equal employment coordination. Another difficulty with this option might well be that the commission's relationships with other agencies have seriously deteriorated over the years.[25] Since federal agencies have often been unable to evaluate critically their own enforcement programs, commission reports have regularly discovered myriad flaws and have candidly criticized the agencies accordingly.[26] Without a far different mandate, the commission would have no direct authority over the equal employment enforcement activities of these agencies, and therefore suggestions on coordination might go completely unheeded or at least ignored in practice.

Consolidation

In view of the disadvantages of coordination options, it is understandable that President Carter turned to the idea of consolidating existing equal employment programs by proposing to Congress what was ultimately called Reorganization Plan No. 1 of 1978. At the outset it is necessary to draw a distinction between consolidation accomplished through the administration's reorganization plan and the use of the normal legislative process to institutionalize reform. Obviously they are related since both approaches require legislative consideration. But it may be that only limited improvements in EEO enforcement will be possible through the president's reorganization plan, and clearly it does not relate to housing or educational discrimination. Carter has realigned equal employment activities within agencies and shifted some of them to EEOC. However, he could not create or abolish agencies through his reorganization plan. Nor could he abolish a function now mandated by statute. Similarly, the options that entail new legislation could generally not be achieved short of the normal legislative process because they would require the abolition and creation of agencies, and they would grant additional powers to the new enforcement agency or agencies.

Administrative consolidation therefore has definite advantages over new legislation: it allows the president substantial leeway without having to proceed through the normal legislative process and without encountering political conflict with Congress over the precise form that equal employment reorganization should take. Equal employment enforcement also may be enhanced if diverse responsibilities in particular issue areas are consolidated into one agency, such as EEOC, that has developed a vigorous enforcement reputation and valuable experience. Nevertheless, President Carter could have reorganized equal employment programs quite differently than he did. Partial and less drastic consolidation alternatives were definitely available to the administration.

The crucial question was which employment-related agency should be given increased responsibility, and President Carter's reorganization task force soon began to favor a Super-EEOC.[27] Carter had the legal authority under his reorganization power to shift equal employment responsibilities from the Department of Labor, the Civil Service Commission, and the Equal Employment Opportunity Coordinating Council to EEOC. The chief shortcoming of this approach is that EEOC has been plagued by major organizational and complaint backlog problems of its own.[28] One important policy trade-off was thus whether to give more authority to an agency fully dedicated to its mission on the one hand, and at the same time impose EEOC's organizational and backlog problems on a Super-EEOC.

Some of EEOC's past problems have been examined by Charles S. Bullock III, one of the nation's foremost authorities on civil rights. Bullock found that

EEOC primarily assumed a passive stance in enforcing equal employment opportunity laws by devoting most of its resources to processing complaints rather than actively investigating job discrimination in major industries. "Not until 1974 did [EEOC] . . . begin to fashion *voluntary* nondiscrimination plans for entire industries."[29] Bullock also found that relatively few complaints were resolved in favor of the complainant, that the backlog dictated that complaints were typically disposed of in two years rather than the sixty days that EEOC set as its goal, that there was a high turnover rate among EEOC's top officials which disrupted the functioning of the agency, and that EEOC investigations were frequently so inadequate that a conviction for job discrimination could not usually be sustained.[30]

Regardless of EEOC's past problems, when the final decision had to be made, probably the most important consideration was the fact that equal employment consolidation would bypass the political problems that would have emerged had the normal legislative process been used to create a new equal employment agency. An attempt to establish a new equal employment agency, rather than a Super-EEOC, would have involved more far-reaching reorganization than consolidation, thereby entailing greater institutional change and perhaps more expenditure of funds over time. The legislative route would also have been necessarily more technical and would have taken longer than reform through executive coordination or the consolidation plan. Diverse views and heated arguments would have arisen with the alteration of existing structures and the modification of present equal employment laws. Some civil rights advocates may have feared the possibility of a coalition of conservative Democrats and Republicans whose objective might have been to weaken EEO rather than to strengthen it. And, as always, political opposition would have arisen to confront any legislative proposal creating a new bureaucracy, even though the administration's obvious intention was to reduce the number and to consolidate the functions of different employment-related agencies. Instead of contributing to the growth of bureaucracy, both the consolidation and legislative approaches would have brought together equal employment professionals from different federal agencies and cut overhead, thereby reducing the size of bureaucracy. For this reason bureaucratic opposition would have inevitably developed. The desire for bureaucratic self-preservation dictates that agencies previously having equal employment authority would be reluctant to give it up, regardless of the improvements that may have resulted from a major reorganization. Some supporters of reform, including those in the civil rights reorganization task force, may therefore have viewed new legislation as an ideal option but supported a Super-EEOC through a reorganization plan in order to avoid apparent political obstacles.

New Legislation

These political and administrative problems would have quickly become evident, for example, had the Carter Administration adopted the U.S. Commission on

Civil Rights' 1975 recommendation regarding the creation of a National Employment Rights Board (NERB). NERB was viewed to be an agency with seven members appointed by the president and confirmed by the Senate for six-year terms.[31] Creation of NERB would have required a new statute modifying several current civil rights laws. A recommendation was made to give NERB sweeping powers, including authority to initiate enforcement procedures; cease and desist authority; power to bring suit in federal district court and to intervene in private actions; authority to debar federal contractors and subcontractors, to terminate federal grants, to decertify labor unions, and to revoke federal licenses; and various investigative powers including subpoena and data reporting. NERB's jurisdiction was to reach only employment discrimination, but it would have been devoted to combating job discrimination on grounds of race, color, religion, sex, national origin, age, and handicapped status. The Commission on Civil Rights estimated that NERB would require appropriations equal to one and one half times those allocated in 1975 to enforce all equal employment laws and regulations. In light of President Carter's stated preference for fiscal conservatism, it was apparent that he would be a reluctant supporter of NERB.[32]

This being so, there were at least two other basic legislative options for the Carter Administration to weigh. The first was to merge all federal civil rights programs into one cabinet-level agency. This would have meant creating a Department of Equal Opportunity. Such a department would necessarily have been given powers greater than those recommended for a National Employment Rights Board. It would have been particularly crucial that the department have the ultimate authority to order other departments and agencies to take specific steps to correct any negative civil rights impacts of their programs. An "equal employment impact statement" might have been required of all federal programs. With such powers the department would have been forced to maintain cordial relationships with other agencies. Were relationships to deteriorate, access to the president would have been necessary to resolve conflicts between the department and other agencies.

A cabinet-level civil rights department would have had obvious advantages. Being solely concerned with civil rights, it could have approached equal employment from a comprehensive and uniform perspective, taking into account the relationships between employment and other forms of discrimination and attacking them accordingly. Under this concept duplicative programs would have been abolished and the nagging problem of redundancy might have finally been solved. It may have even been possible for a departmental structure to reduce the total amount of public funds previously devoted to myriad equal employment programs.

On the other hand, the disadvantages might have been overwhelming. Even assuming that political opposition was overcome and that the department was established, this option would have caused unprecedented shakeups in the area of equal employment and would have destroyed existing contacts between civil rights groups and individual agencies. It could have caused rivalries between existing agencies due to their loss of equal employment programs and thereby contributed to a lack of future interdepartmental cooperation. Where conflicts developed between the department and other agencies, presidential intervention would have been essential. The department clearly would have been in trouble in the case of a president who held civil rights as a low priority. Moreover, by placing all civil rights enforcement activities into one department, a future president—one not supportive of civil rights—could more easily have curbed programs and reduced the budget of a single agency, rather than the many that had existed.

The final option proffered here would be a civil rights agency independent of the executive branch, similar to most regulatory agencies. An independent enforcement agency would have powers equivalent to those of a department but would have been more insulated from political pressures. Although the head or heads of such an agency would be appointed by the president and approved by the Senate for a fixed term, the agency would possess more control over its resources and policies than is the case with other options. The disadvantages of creating an independent agency are similar to those associated with a department, but they would very likely be even greater. Both the president and Congress would tend to oppose establishing a civil rights agency with such power and autonomy.

The Aftermath: EEOC and Reorganization

Demonstrating his leadership qualities and his concern for EEO, President Carter has made some outstanding appointments to high-level EEO positions, such as Eleanor Holmes Norton to chair EEOC. We will now examine how Norton and other leaders at EEOC have internally reorganized the agency to resolve employment discrimination cases more efficiently and, second, how the agency has planned to perform its new responsibilities that were transferred to it from the Civil Service Commission, the Department of Labor, and the Equal Employment Opportunity Coordinating Council.[33]

During 1978 EEOC underwent a major structural and procedural reorganization. According to the commission, all its processes to handle employment discrimination complaints were "redesigned into charge processing systems geared to increase efficiency and speed."[34] Three model offices were established in Baltimore, Chicago, and Dallas. Their assignment was to process old and new employment discrimination complaints rapidly, to negotiate settlements quickly, and "for the first time [to have] lawyers . . . working with investigators in an integration of the investigation, conciliation and litigation functions."[35] The three model offices worked so productively with their new Rapid Charge Processing System that the system was set up in all EEOC district offices in 1978.

There are two principal components of this Rapid Charge Processing System that should be underscored. First is detailed interviews with a complainant as soon as the complaint is filed. Second, a face-to-face fact-finding conference between the complainant and the respondent is scheduled "in a few weeks after the filing of the charge."[36] During fact-finding conferences, evidence from all parties is promptly gathered, and an early attempt is made to resolve the dispute. Where resolutions are not forthcoming at this stage, an additional investigation is scheduled with the hope of conciliation. Typically, an employer's work-force profile is closely examined, and if minorities and women are underrepresented, EEOC often offers to help the employer establish a remedial program. If conciliation is still unattainable, even after technical assistance from Washington headquarters, EEOC then initiates a lawsuit in federal district court if the evidence points toward discrimination. After the lawsuit, the employer's hiring and promotion practices are strictly monitored by EEOC's district offices.[37]

In addition to its Rapid Charge Processing, EEOC has developed a new Backlog Charge Processing System and a program known as Early Litigation Identification. With regard to the former, EEOC claims that "[for] the first time in its history, the Commission [is able to] premanently [reduce] its backlog" through the Backlog Charge Processing System.[38] This of course suggests that instead of EEOC's past dilemma of having to deal constantly with a huge backlog of complaints, the agency can focus most of its attention upon complaints as they are filed. In October 1979, EEOC predicted that its entire backlog would be eliminated within two years.[39] The Rapid Charge Processing System and the Backlog Charge Processing System are now installed in thirty-seven new area EEOC Offices that were created "to make the Commission more accessible to the public and [to] facilitate rapid processing of cases."[40] These systems are also expected to provide EEOC with enough time to target in on smaller companies, as opposed to just large employers.[41] The Early Litigation Identification program is in fact primarily aimed at smaller companies and is intended to increase the number of class action suits filed by EEOC field offices.[42]

At the same time that field offices were being reorganized, EEOC was simultaneously changing the structure of its headquarters offices in Washington. "The new headquarters reorganization has been designed to better serve the agency's field offices" and includes an office for Field Services, Systemic Programs, Management and Finance, Policy Implementation, Government Employment, Special Projects and Programs, Public Affairs, and Administrative Service.[43] This headquarters reorganization is aimed at eliminating what had previously been highly overlapping office structures and functions in Washington.

When considered as a whole, EEOC's internal reorganization and new procedures won full support and approval from Chairman Augustus F. Hawkins of the House Subcommittee on Employment Opportunities. According to Congressman Hawkins, Chairperson Norton "is to be commended for making very real progress in reducing the backlog of charges filed with [EEOC] and for taking strong leadership in establishing the Commission as a competent, professional law enforcement agency."[44] In reply, Norton noted that "major reforms affecting every major system in the agency [were] either complete or well underway."[45]

Now that EEOC has been internally reorganized, how is the agency planning to carry out its new responsibilities earlier housed in the Civil Service Commission, the Department of Labor, and the Equal Employment Opportunity Coordinating Council? What follows is a preliminary answer to the question. It may indeed be too soon to get an accurate picture of EEOC's efficiency and effectiveness in executing its new duties, for these have largely been undertaken only in 1979. Nevertheless, a preliminary assessment of EEOC's plans to take over these broader duties is both worthwhile and timely.

According to Eleanor Holmes Norton, the EEO coordination function that was theoretically but never adequately performed by the Equal Employment Opportunity Coordinating Council was assumed by EEOC as early as July 1978 and is now in operating order. In late November 1978, she noted that the coordination function was then in effect even though it lacked sufficient resources and staff. Norton explained that since the coordination function was so important, EEOC initially established a Task Force on Interagency Coordination, composed of officials from EEOC's headquarters office. The task force then developed a new office at EEOC, known as the Office of Interagency Coordination (OIC). Like the task force, OIC is operated by EEOC staff from other parts of the agency. OIC's chief job is working with other federal agencies that request help in reaching conciliation in EEO cases. In November 1978, the task force was dismantled and an acting director was named for OIC. Thus, Chairperson Norton concluded that "OIC is now operational as a Commission office using a small staff composed of EEOC employees detailed to further develop and discharge coordination responsibilities. In the continued absence of budget authority for these functions, some OIC matters are also referred to EEOC headquarters staff."[46]

Regarding the transfer of the Civil Service Commission's jurisdiction over EEO in the federal government and the Department of Labor's responsibility to enforce the Equal Pay Act and the Age Discrimination in Employment Act, Norton commented that EEOC did not assume those responsibilities until January 1 and July 1, 1979, respectively. However, in early 1978 EEOC was already preparing to take over these functions, even before Congress approved Reorganization Plan No. 1 in May 1978. Although EEOC had not actually assumed the EEO duties of the Civil Service Commission and the Department of Labor, in late 1978 Norton outlined to the House Subcommittee on Employment Opportunities the planning process that had developed.[47]

First, EEOC had exchanged employment data and staffing patterns with the Civil Service Commission and the Department of Labor in anticipation that those agencies would be required to transfer some of their personnel to EEOC after the reorganization was implemented. Second, EEOC developed in advance a recruitment and training program for personnel transferred from both agencies, based on the assumption that EEOC's EEO standards and programs would differ significantly from those of the other two agencies. Third, EEOC "approved the appeals structure for the CSC function. Appeals procedures for the CSC function [were] ready for Commission scrutiny and approval. The procedures also [addressed] the over 2,400 case backlog now at the CSC."[48] Fourth, EEOC plans to retain the existing procedures of the Civil Service Commission and the Department of Labor for up to one year. EEOC nevertheless has begun developing overall plans to reform these old procedures. "After detailed study of the procedures in their application, and consultation with protected groups, unions, and business, changes will be implemented."[49] Fifth, EEOC encouraged the Civil Service Commission and the Department of Labor to detail in advance some staff for planning new EEOC activities previously held by the other two agencies. Finally, Norton had met about every three months with minority and female group representatives to receive input on EEOC's exercise of its new responsibilities and how those responsibilities could best be carried out to benefit the protected groups.

Through Executive Order 12106, President Carter officially made EEOC responsible for eliminating discrimination in the federal employment as of January 1, 1979. All complaints filed by federal employees under title VII of the Civil Rights Act of 1964 now go directly to EEOC. As of early 1979, EEOC was still using essentially the same procedures previously developed by the Civil Service Commission. "Under these procedures, EEOC will hold a hearing on the complaint which, if rejected, may be appealed to EEOC's newly established Office of Appeals and Review. Should the employee reject EEOC's final decision, he then may take his case to court."[50]

EEOC's planning procedures, all taken in advance of its assumption of new authority, and subsequent actions pursuant to Executive Order 12106, suggest that the commission is prepared to carry out its new duties. The Rapid Charge Processing System and the Backlog Charge Processing System also lead one to believe that EEOC is able to handle the heavy caseload formerly under the jurisdiction of the Civil Service Commission and the Department of Labor. However, again it must be emphasized that this is a preliminary assessment. It has only been since January 1979 that EEOC actually began to oversee job-discrimination complaints in the federal government and July 1979 that it began enforcing the Equal Pay Act and the Age Discrimination in Employment Act. Since recent information on the results and payoff of Reorganization Plan No. 1 is difficult to come by, one can merely hope at this point that EEOC's early planning was adequate enough for it to execute its new duties efficiently and effectively.

Conclusions

Achieving compliance with the nation's equal employment laws has been difficult, to say the least. These laws require substantial changes in much of the political, social, and economic status quo—changes in who gets what, when, and how.[51] Issues such as affirmative action in employment are extremely controversial. Experience has shown that EEO cannot be brought about through an ad hoc process of individual complaint investigations by a variety of federal agencies. But in the final analysis, compliance can be significantly enhanced through the enforcement efforts of a consolidated civil rights agency with a desire to eliminate job discrimination and with sanctions applied firmly and fairly. This is apparently why President Carter chose to create a Super-EEOC in Reorganization Plan No. 1, even though he had a wide range of other policy options.

At this early date it appears that Reorganization Plan No. 1 is being carried out reasonably well by EEOC. Advance planning was the key. Prior to taking over EEO responsibilities from the Civil Service Commission and the Department of Labor in 1979, EEOC had established an agenda for reform. It had worked with the Civil Service Commission, the Department of Labor, and protected minority groups and women to anticipate needed changes in EEO enforcement policy. In addition, EEOC leaders have internally reorganized the agency, developing systems that wiped out many backlog cases and that better ensured that new EEO complaints could be handled rapidly as they were filed. While it is admittedly too soon to conclude that President Carter's reorganization of equal employment enforcement will be an overwhelming success, it is clear that he provided the presidential leadership necessary to promote that end, that he selected one of the least politically vulnerable of the many competing policy options for EEO reorganization, and that his spirited leadership carried over to EEOC in its internal reorganization and its assumption of new equal employment responsibilities.

Notes

1. See Jack Greenberg, *Race Relations and American Law* (New York: Columbia University Press, 1959), ch. 6.

2. For a relatively recent account see Donald J. McCrone and Richard J. Hardy, "Civil Rights Policies and Achievement of Racial Economic Equality, 1948-1975," *American Journal of Political Science* 22 (February 1978): 1-17. For a slightly dated but still a reasonably accurate picture of employment discrimination see Charles S. Bullock III and Harrell R. Rodgers, Jr., *Racial Equality in America: In Search of an Unfulfilled Goal* (Pacific Palisades, Calif.: Goodyear Publishing Company, 1975), ch. 4.

3. Article II, section 3 of the Constitution requires that the president "shall take Care that the Laws be faithfully executed, . . ."

4. Richard P. Longaker, *The Presidency and Individual Liberties* (Ithaca, N.Y.: Cornell University Press, 1961), p. 196. See also M. Glenn Abernathy,

Civil Liberties under the Constitution, 3d ed. (New York: Dodd, Mead, and Company, 1977), pp. 91-94.

5. U.S. Commission on Civil Rights, *The Federal Civil Rights Enforcement Effort–A Reassessment* (Washington: U.S. Government Printing Office, 1973), p. 11.

6. Martin Tolchin, "President Proposes Merger of Programs in Fight on Job Bias," *The New York Times,* February 24, 1978, p. 1. Eleanor Holmes Norton similarly stated that President Carter's reorganization plan "presents an unprecedented opportunity for the EEOC to lead a revitalized, consistent and effective Federal equal employment effort." U.S., Congress, House, *Oversight on Federal Enforcement of Equal Employment Opportunity. Hearings before the Subcommittee on Employment Opportunities of the House Committee on Education and Labor,* 95th Cong., 1st sess., 1979, p. 41.

7. "Reorganization Plan No. 1 of 1978," *Federal Register* 43 (May 9, 1978): 19807-19809. See also *Weekly Compilation of Presidential Documents* (January 1, 1979): 2290-2292. The reorganization plan was implemented by EEOC according to a staggered schedule.

8. *Weekly Compilation of Presidential Documents* (February 9, 1977): 170.

9. The Reorganization Act of 1977, P.L. 95-17. For a general discussion of the reorganization statute and the Carter Administration's actions pursuant to it, see John P. Plumlee, "Carter's Major Structural Reorganizations: An Appraisal" (Paper presented at the 1979 Annual Meeting of the Midwest Political Science Association, April 19-21, 1979).

10. "Howard Glickstein Appointed to Head OMB Task Force on EEO Reorganization," *Daily Labor Report* (May 20, 1977): 5. In retrospect it is interesting to note that in April 1977 EEOC Commissioner Daniel Leach recommended that his agency be given all equal employment enforcement responsibilities. Leach's proposal presumably would not have required new legislation since it would have only reassembled authorized equal employment activities of various agencies and would not have given the central employment enforcement agency any additional legal powers to eliminate job discrimination. Leach also believed that coordination was the major problem facing Carter's reorganization effort. He stated that "no less than seven major pieces of legislation and executive orders impose equal employment opportunity obligations" and that "this multiple agency approach, rather than strengthening enforcement, weakens it. It is duplicative, confusing, and inconsistent." "EEOC Commissioner Daniel Leach Calls for Merger of Federal Civil Rights Agencies into Super-EEOC," *Daily Labor Report* (April 18, 1977): 11-12.

11. "OMB Task Force on Civil Rights Reorganization Plans to Complete Report by Mid-October," *Daily Labor Report* (June 3, 1977): 5.

12. "OMB Task Force on EEO Underway," *Daily Labor Report* (June 3, 1977): 2.

13. Ibid.; "OMB Task Force on Civil Rights," p. 6.

14. U.S. Commission on Civil Rights, *The Federal Civil Rights Enforcement*

Effort: To Preserve, Protect and Defend the Constitution (Washington: U.S. Government Printing Office, 1977), pp. 25-27, 64-66.

15. U.S. Commission on Civil Rights, *The Federal Civil Rights Enforcement Effort: To Eliminate Employment Discrimination: A Sequel* (Washington: U.S. Government Printing Office, 1977), pp. 329-330.

16. See generally Charles O. Jones, *An Introduction to the Study of Public Policy,* 2d ed. (North Scituate, Mass.: Duxbury Press, 1977), pp. 147-150.

17. Advisory Commission on Intergovernmental Relations, *Improving Federal Grants Management* (Washington: U.S. Government Printing Office, 1977), p. 81.

18. See John Hope II, *Minority Access to Federal Grants: The Gap between Policy and Performance* (New York: Praeger, 1976), ch. 2.

19. Commission on Civil Rights, *Defend the Constitution,* p. 189.

20. Comment, "Title VI of the Civil Rights Act of 1964–Implementation and Impact," *George Washington Law Review* 36 (May 1968): 857-860.

21. Ibid., pp. 875-876.

22. Charles M. Lamb, "Administrative Coordination in Civil Rights Enforcement: A Regional Approach," *Vanderbilt Law Review* 31 (May 1978): 864-876.

23. Ibid., pp. 859-860, note 24.

24. "Title VI," p. 868.

25. See Robert S. Rankin, "The Civil Rights Movement from the Vantage Point of the Civil Rights Commission," *Oklahoma Law Review* 25 (February 1972): 105.

26. Charles S. Bullock III and Harrell R. Rodgers, Jr., "Impediments to Policy Formulation: Perception Distortion and Agency Loyalty," *Social Science Quarterly* 57 (December 1976): 506-519.

27. See the suggestion of EEOC Commissioner Leach, note 13.

28. See Harrell R. Rodgers, Jr., and Charles S. Bullock III, *Law and Social Change: Civil Rights Laws and Their Consequences* (New York: McGraw-Hill, 1972), pp. 119-122. For general assessments of EEOC's past successes and failures see Commission on Civil Rights, *Employment Discrimination: A Sequel,* ch. 4; idem, *The Federal Civil Rights Enforcement Effort: To Eliminate Employment Discrimination* (Washington: U.S. Government Printing Office, 1975), ch. 5.

29. Charles S. Bullock III, "Expanding Black Economic Rights," in Harrell R. Rodgers, Jr., ed., *Racism and Inequality: The Policy Alternatives* (San Francisco: W.H. Freeman Company, 1975), p. 81 (emphasis added).

30. Ibid., pp. 81-85.

31. Commission on Civil Rights, *Employment Discrimination,* pp. 651-654.

32. Given President Carter's leadership in equal employment opportunity, the Commission on Civil Rights did not again recommend the specific creation of NERB in its 1977 equal employment report. Nevertheless, the commission adhered to its basic position that there should be the "creation of a single

agency, enforcing equal employment opportunity under a single law." Commission on Civil Rights, *Employment Discrimination: A Sequel,* p. 333.

33. Developments discussed in this section cover the period from January 1978 through October 1979.

34. *EEOC: The Transformation of an Agency* (Washington: Equal Employment Opportunity Commission, 1978), p. 1. The commission's reorganization steps are discussed in detail in "Equal Employment Opportunity Commission: Procedural Regulations," *Federal Register* 44 (January 23, 1979): 4667–4670; U.S., Congress, *Oversight on Federal Enforcement,* pp. 4-82.

35. *EEOC: Transformation of an Agency.* p. 1.

36. Ibid, p. 2.

37. Ibid., p. 3. Using these procedures, the three model offices proved to be clear success stories. In November 1978 Norton noted that "[i]n the model offices, new charges are being closed, on the average, in 65 days. Settlements and conciliations have increased from 14 percent under the old system in fiscal year 1977, to 48 percent under the new system in fiscal year 1978." U.S., Congress, *Oversight on Federal Enforcement,* p. 6.

38. *EEOC: Transformation of an Agency,* p. 4.

39. "Litigation Backlog Trimmed, EEOC Says," *Daily Labor Report* (October 16, 1979): 2.

40. *EEOC: Transformation of an Agency,* p. 5. Ten of these thirty-seven smaller offices were previously district offices that had been full-service units prior to EEOC's internal reorganization. Instead of being concerned with agency backlogs as before, most of these ten area offices contain only the Rapid Charge Processing System. These reforms were effective as of January 29, 1979. "Equal Employment Opportunity Commission," p. 4668. For a more detailed discussion see "10 Major Cities Will Lose Clout under EEOC Field Reorganization," *Daily Labor Report* (January 24, 1979): 4-5.

41. "EEOC to Target Smaller Firms," *Daily Labor Report* (February 8, 1979): 1, 4.

42. "EEOC Commissioners Approve Plan for Expanded Systemic Actions," *Daily Labor Report* (February 8, 1979): 4-5.

43. *EEOC: Transformation of an Agency,* p. 8.

44. U.S., Congress, *Oversight on Federal Enforcement,* p. 1.

45. Ibid., p. 2.

46. Ibid., pp. 42–43.

47. Ibid., pp. 44–45.

48. Ibid., p. 44.

49. Ibid. In July 1979 EEOC announced that it would issue its own guidelines later in 1979 involving the Equal Pay Act and the Age Discrimination in Employment Act. See "EEOC Will Not Adopt Labor Department's Standards for Age Discrimination in Employment, Equal Pay Acts," *Daily Labor Report* (July 11, 1979): 5-7.

50. "Executive Order Transfers EEO Functions of Civil Service Commission to EEOC," *Daily Labor Report* (January 3, 1979): 5.

51. An obvious recent example, of course, is the Supreme Court's interpretation of title VII of the Civil Rights Act of 1964 in Weber v. Kaiser Aluminum & Chemical Corp., 99 Sup. Ct. 2721 (1979).

7 *Bakke, Weber,* and Race in Employment: Analysis of Informed Opinion

J. David Gillespie and
Michael L. Mitchell

American business and labor displayed an extraordinary interest in *Regents of the University of California* v. *Bakke* 438 U.S. 265, 98 S.Ct. 2733, 57 L.Ed. 2d750 (1978). Among sixty-one amici, the most ever filed in a Supreme Court case, there were briefs from the U.S. Chamber of Commerce, American Subcontractors Association, four federations of the AFL-CIO, the United Auto Workers, United Mine Workers, and other associations representing business and labor interests.

In June 1978, a divided court ruled that Allan Bakke, a white, had been illegally denied admission to medical school at the Davis campus of the University of California. The Davis medical school had established for its admissions process an affirmative action program which had the effect of reserving 16 of the 100 freshman seats for minority applicants, even though the relatively new medical institution bore no prior record of discrimination against minorities. The Supreme Court case yielded not one written opinion but six. Justice Lewis Powell constructed the "opinion of the court." With Burger, Stevens, Rehnquist, and Stewart, Powell voted for the admission of Bakke and ruled that racial quotas in universities and other federally funded programs violate title VI of the 1964 Civil Rights Act. At the same time, Powell agreed with Brennan, White, Marshall, and Blackmun that race may be an acceptable factor for consideration in an admissions process if such consideration serves a defined and legitimate purpose.[1]

Although the Supreme Court might have chosen to base its decision on the "equal protection" clause of the Fourteenth Amendment, it invoked instead a statutory foundation: title VI, which prohibits racial and ethnic discrimination in federally funded programs. Parallel to title VI in the 1964 Civil Rights Act is title VII, banning private employment discrimination based on race, color, sex, religion, or national origin.

Business and labor interest in the *Bakke* case was related to the intimate substantive linkage between the two titles of the 1964 act. In 1972 the protection afforded by title VII to private-sector employees was extended to public employees. Present federal legislation also prohibits racial, sex, ethnic, or religious discrimination by unions.[2]

Federal equal employment statutes are administered by the Equal Employment Opportunity Commission (EEOC). From its inception, EEOC has been authorized to investigate charges of employment discrimination; but only with the passage of the 1972 Equal Employment Opportunity Act was it empowered to initiate federal court action.[3]

In 1971 the Supreme Court, in the landmark *Griggs* v. *Duke Power Company,* unanimously ruled that when job qualifications and tests have an adverse racial effect (for example, reducing the number of successful minority applicants), they may be deemed discriminatory, even if discrimination is not the intent, unless they can be verified as accurate indicators of job performance.[4] In general, federal courts have been friendly to EEOC efforts to combat job discrimination; for example, ruling that when percentages of minority or female employees are far below those in the community or in the available work force, the burden of proof rests with the employer who denies discrimination.[5]

For some years, federal courts have required business firms to take affirmative action, often including reinstatement and remuneration, in individual cases of proven discrimination. The EEOC has imposed upon particular employers steps involving numerical goals and timetables that closely resemble quotas. Judicial doctrine before 1979 was that quotas were illegal except in verified cases of prior discrimination; some, following *Bakke,* questioned the legality of quotas under any circumstance.[6]

One of the problems confronting EEOC is its complaint caseload, the heavy volume of which has left the agency sometimes as much as two years behind schedule. For this reason, EEOC has counted upon "voluntary" measures taken by employers to avoid eventual agency action. In this, however, employers often faced a dilemma: in voluntarily establishing any program involving what amounted to quotas, employers were probably within legal bounds only if they admitted past discrimination. But such an admission might subject them to suits by present and former employees seeking remuneration as victims of the conceded discrimination. Not unexpectedly, many companies chose inactivity, or at least avoided the type of affirmative action proven most effective in yielding numerical results.[7]

In December 1978, the Supreme Court agreed to hear the cases *Kaiser Aluminum and Chemical Company* v. *Weber* and *United Steelworkers of America* v. *Weber,* thereby committing itself to face squarely the issue of affirmative action, including quotas, in the employment field.[8] Kaiser, under an agreement with the United Steelworkers, established at its Gramercy, Louisiana, plant a training program to prepare trainees for skilled craft jobs. The program was initiated in part to increase the number of black craft employees at the plant, where in 1974 blacks had held only 5 of 290 skilled jobs. Accordingly, it admitted one black for every white selected. Brian Weber, a white, was denied a place in the training program even though his seniority at the plant was greater than that of most of the blacks admitted.[9]

The Supreme Court, in a June 1979 *Weber* decision, reversed the rulings of two lower federal courts that Kaiser, in the absence of any proven or admitted record of discrimination against blacks, violated title VII by preferring blacks in its training program. Stewart, White, Marshall, and Blackmun joined in the opinion, written by William J. Brennan, Jr., that voluntary affirmative action plans, including (within limits unspecified in the decision) the use of quotas such as Kaiser's, are allowed under title VII. Rehnquist, joined by Burger, issued an extraordinarily acrimonious dissent.[10]

There were several parallels between *Weber* and *Bakke:* affirmative action programs involving what amounted to quotas, initiated by institutions not found guilty of prior discrimination against minorities, and challenged under specifications of the 1964 Civil Rights Act. Of the two decisions, *Weber* represents an unambiguous victory for affirmative action. Though *Bakke* was less favorable to expressed minority interests, the strictures of that decision have been sometimes overstated.

Advocates of affirmative action in the employment field were in fact relieved at the restraint shown by the court in *Bakke,* especially in its not expressly applying the restrictive principles to private-sector-employment practice. The Supreme Court itself seemed to underscore this caution in subsequent decisions. In July 1978, it let stand a consent decree requiring American Telephone and Telegraph to promote more female and minority employees and to pay victims of past discrimination $42 million in back wages and raises. And it refused to hear a challenge to the 1977 Public Works Employment Act provision earmarking 10 percent of dispersed grant monies under the act for minority businesses.[11]

Shortly after the announcement of the *Bakke* decision, Eleanor Holmes Norton, EEOC head, warned private employers that they "proceed at great risks if they use *Bakke* as an occasion for retreat from their obligations under title VII." Said Norton, "we believe numerical remedies, including quotas under certain circumstances, are still permissible and we will not stop them until the courts tell us we must."[12]

Still, the *Bakke* decision left an aura of uncertainty, the more because of the divisions within the court itself. Some observers predicted that the decision would have a direct impact upon employment, probably more upon employer and minority attitudes and actions than upon public policy and administration. Perhaps a larger number perceived the *Bakke* decision as an important presage of employment policy changes in judicial decisions to be made. It was in the context of this sense of uncertainty and significance that the present study of informed opinion about affirmative action in employment was made.[13]

The Study and Findings

Methodology. The study was made using Q methodology. Q is unlike probability sampling surveys, in which inferences are drawn about views of a population based upon structured responses of a sample to discrete questions. Q methodology allows the respondent to create and register a model of his attitude by relating various responses to each other. Q methodology "is an approach that determines the major points of agreement and disagreement—and their relative significance—in the population by analyzing the responses of a rather small number of individuals, selected to represent the major perspectives on an issue, to a set of statements chosen to cover a wide range of viewpoints on the subject (the Q-sample). Each respondent provides a Q-sort, a ranking of his agreement or disagreement with the statements. The Q-sorts are then correlated and factor analyzed to isolate the various common attitudes."[14]

In the present study, the authors assembled a Q-sample composed of fifty-six statements concerning *Bakke* and employment practice. These statements were drawn from many sources: amici briefs, news magazine articles, and informal "street interviews." Most of the fifty-six fit into one of three categories: normative or value statements; statements treating the direct impact of *Bakke* upon employment practice; and statements treating *Bakke* as a portent of things to come.

Respondents were asked to read carefully through the statements printed on fifty-six individual cards, and then to sort the cards into three piles: agree, disagree, and neutral (figure 7-1). Eventually each respondent placed the number representing each statement at the position he deemed appropriate on a score sheet. The numbers corresponding to the three statements with which the individual respondent was in greatest disagreement were placed in the –5 column. The three most agreeable statements were to have their numbers placed in the +5 column; and so on.

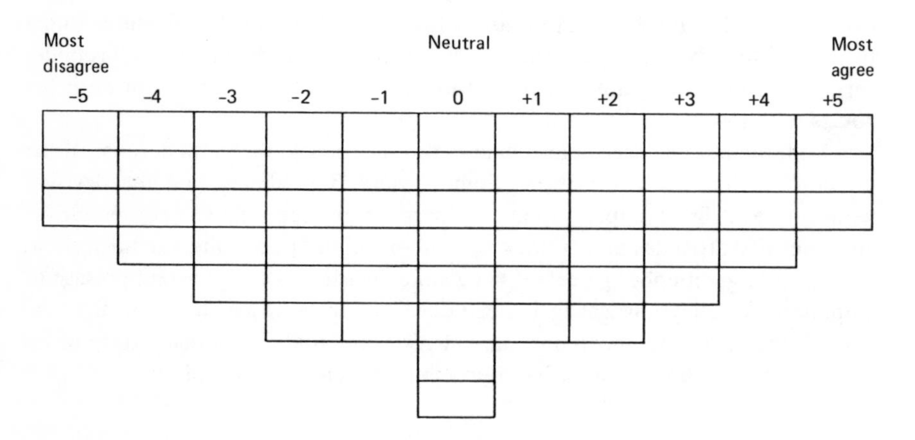

Figure 7-1. Sample Score Sheet

The focus of the inquiry was upon *informed* opinion. Requests for responses were thus sent to government agencies, groups, and associations–educational interests, civil rights and minority groups, professional associations, corporate and trade associations, unions, and ideological groups, among others–presumed interested, many having filed amici briefs, in the *Bakke* case.

Responses were made in October and November 1978. Returns based upon the mailings were considerable, in view of the nearly two hours it took for the average respondent to complete the researchers' request. The data include responses from a local Chamber of Commerce, the American Civil Liberties Union, Young Americans for Freedom, Yale Black Law Students Union, Mexican-American Legal Defense Fund, Pacific Legal Foundation, American Jewish Committee, Los Angeles County Bar Association, National Council of La Raza, and several state and federal agencies. Spokespersons for groups, associations, and agencies were informed that their names and positions would not be published and that their responses would be treated as individual and not as official expression of group or agency position. Faculty members in law, business, and economics at a southern university and professors and instructors in political science at a midwestern university were also included in the study and its results.

Results. The study revealed among the informed three principal attitudes (represented by Factors I, II, and III) with regard to *Bakke* and affirmative action in employment. The scores given to each of the fifty-six Q-sample statements by each factor, that is, by respondents associated with each factor, are given in Appendix 7A.[16]

Factor I represents an attitude that strongly supports and defends affirmative action, including numerical quotas, in employment practice as in school admissions. It assigns the highest priority to the goals of integrating minorities into the American mainstream; and it regards as both hypocritical and unrealistic, considering the pervasive legacy of historical discrimination against racial minorities, the arguments of those who now espouse employment practices devoid of all race consideration. Race-conscious affirmative action programs are seen by factor I respondents as just in principle, because in American society minority individuals must struggle harder in order to achieve a position of status and responsibility. Thus rejected by factor I is the implication that programs preferring racial minorities forgo serious consideration of applicant qualifications. Factor I normatively assigns to the courts the role of protecting minority rights, considering that the elected branches are likely to respond to majority preference. And it was very critical of the *Bakke* decision, foreseeing an adverse impact upon the attitudes and actions of minority individuals and employers and upon federal and state affirmative action enforcement efforts. The decision, said factor I respondents, sowed the seeds of doubt and uncertainty. And from *Bakke,* they forecast–with a pessimism semmingly undue, now considering the subsequent *Weber* decision—that there would be adverse rulings in future employment cases.

The correlation between factors I and II is –0.083 (range: –1.0 to +1.0), which is insignificant. The factor II attitude perceived *Bakke* as a manifestation of a conservative trend in American life and values, and found in the case greater significance than did factors I or III. Factor II agrees that nonwhites have had to struggle harder; and it concedes (+1) that bringing the minority into the American mainstream must be a goal of highest priority. But it condemns as unjustified and unnecessary the use of quotas or indeed of any race-sensitive decision procedure. Factor II contends that qualification, not race, should determine who is admitted to a school or job. Respondents associated with factor II hold that the equal protection clause must be reinterpreted to protect whites as well as nonwhites, individual victims irrespective of racial category. The factor II attitude is in accord with certain aspects of the *Bakke* decision. Its respondents foresaw, presumably with pleasure, a move by employers away from voluntary programs, and increased difficulties in enforcement by state and federal agencies. But they denied that *Bakke* would affect negatively either the psychological or the material circumstances of nonwhites. And they challenged the Supreme Court to be consistent in post-*Bakke* decisions (such as *Weber*): factor II affirmed (+4) the statement (38) that "it would be inequitable and inconsistent for the Supreme Court, after providing relief for Bakke, to deny the same to a blue-collar white denied employment or promotion as a result of circumstances virtually identical to those pertaining to Bakke's rejection by the medical school." But factor II is strongly critical of the authority given by the Supreme Court (in *Bakke*) to certain race-sensitive programs. Alone among the three principal factors, it affirms the proposition that as a result of the judicial authorization of such programs, whites now do not possess even the limited constitutional protection held by blacks under the former "separate but equal" doctrine.

Factor III is positively but insignificantly correlated with factor I (+0.087). The factor III correlation with factor II is also positive and without much significance (+0.294). Factor III represents an attitude that is more functionalist than ideological. It holds that the minority has had to struggle harder, and it emphasizes the objective of social integration of nonwhites. But it asserts that equity presupposes the extension of constitutional protections for nonwhites to whites as well. Factor III respondents believe it important for employers and admissions officers to take into account qualification; but they strongly reject the contention that judicial deference to race-sensitive programs encourages the mediocre and unqualified. Factor III defends and supports the *Bakke* decision for its balance and moderation. *Bakke* was seen as significant, but *not* in fostering changes in the employment area. The decision, said factor III respondents, would have no adverse effect upon the psychic or material circumstances of nonwhites or the affirmative action efforts of state and federal agencies. The factor

III attitude rejects numerical quotas, but repeatedly affirms and supports the concept of race-sensitive programs. It does maintain some hope for the effectiveness of race-neutral progams that consider instead culture, education, and income.

Policy Implications

Pending before the Supreme Court at the time of this study, and standing in the lineage of *Bakke,* was the *Weber* case. Further back in the judicial process but likewise with a potential for national policy impact was a California appeals court decision overruling an Oakland quota plan wherein minority individuals were given preference to whites in the hiring and promotion of fire fighters. In its ruling, the appeals court averred that the *Bakke* decision was "clear authority that a minority individual quota provision having the effect of discriminating against faultless Caucasians on the basis of race alone is unlawful."[17]

In the American political system, the Supreme Court plays an important, if widely misunderstood, role in policy formulation. Decisions, especially landmark ones, are therefore "political" in substance. Justices venerate the law, but also appreciate the importance of the context within which law is interpreted. Significant also are the political philosophies and social persepctives of various judges. And though the point is always in danger of overstatement, it is true that the Supreme Court is not unmindful of and that it may occasionally be influenced by public opinion on a particular issue. This is attributable in part to the reliance of the Court upon a public perception of the legitimacy of the Court as an institution and of the policy decisions made by it.

Q methodology has the power to present a spectrum of discrete attitudes with respect to a complex issue. It cannot infer the number of people associated with a particular attitude. The present study does reveal that a segment of informed opinion—people holding the attitudes represented by factor I—readily supports the Supreme Court *Weber* decision and others like it. If in the future the Supreme Court should seek to construct in the employment area a *Bakke*-type framework (that is, race-sensitive plans but no quotas), or if it should entirely abrogate the discretionary use of race in employment decisions, it would in either case be supported by some element of informed opinion (factor III and factor II, respectively).

Present majority opinion among the American people appears most congruent with factor III and the middle ground found by the Court in its *Bakke* decision. A *New York Times*/CBS News Survey conducted in October 1977 revealed that among whites 59 percent and among blacks 83 percent favored "special consideration" for well-qualified minority applicants to colleges and graduate schools, but that 60 percent of the whites (42 percent of the blacks) disapproved of minority quotas in such circumstances. Likewise, 63 percent of the whites and 88 percent of the blacks agreed that large companies should be required to set up "special training programs for members of minority groups," while only 35 percent of the whites (64 percent of the blacks) approved of requiring minority hiring quotas.[18]

Public opinion provides a range of tolerance with regard to policy alternatives for any particular issue. As for affirmative action, the center point of that tolerance appears to be upon a policy allowing and promoting affirmative action but excluding quotas. Public perception of affirmative action in employment is permissive, as indicated by the *New York Times*/CBS News data indicating broad support for minority training programs. Controversy over the limited endorsement given by the Supreme Court to quotas in *Weber* will not undermine the authority of the decision or of the affirmative action framework that it embraced and established.

Notes

1. "What the Court Said in Two 5 to 4 Rulings on the Bakke Case," *Chronicle of Higher Education,* July 3, 1978.
2. Charles Bulmer and John L. Carmichael, Jr., "Labor and Employment Policy: An Overview of the Issues," *Policy Studies Journal* 4 (Winter 1977): 258-259.
3. Ibid.
4. Allan P. Sindler, *Bakke, DeFunis, and Minority Admissions* (New York: Longman Inc., 1978), pp. 180-181.
5. "The Tale of Title VII," *Time* 112 (July 10, 1978): 17.
6. Ibid.
7. Owen Ullman, "Factory Worker's Suit May Make Him New Allan Bakke," *Minneapolis Tribune,* July 23, 1978.
8. *U.S. Law Week* 47 (December 12, 1978): 3401-3402.
9. "Bigger Than Bakke? 'Reverse Discrimination' Returns to the Court," *Time* 112 (December 25, 1978): 47; also "Now a 'Bakke' Case for Jobs," *The New York Times,* December 17, 1978.
10. *Weber* was a 5 to 2 decision. Neither Powell, who had been the principal architect of the *Bakke* decision, nor Stevens took part in the *Weber* case.
11. "Bakke: Some Changes," *Newsweek* 92 (July 17, 1978): 29-30.
12. Quoted by Austin Scott for the *Los Angeles Times* and printed in newspapers across the country.
13. This study was supported in part by Grant No. 69, Samford University Research Fund.
14. Steven R. Brown and James G. Coke, *Public Opinion on Land Use Regulation* (Columbus, Ohio: Academy for Contemporary Problems, 1977), p. 3. For a description of Q methodology and its theoretical base, see William

Stephenson, *The Study of Behavior: Q-Technique and Its Methodology*, Midway ed. (Chicago: University of Chicago Press, 1975); also *Operant Subjectivity*, the Q methodology newsletter edited by Steven R. Brown.

15. Philip B. Kurland and Gerhard Casper, eds., *Landmark Briefs and Arguments of the Supreme Court of the United States: Constitutional Law*, 1977 Term Supplement, vols. 99-100 (Washington: University Publications of America, 1978).

16. A fourth factor was also revealed. Factor IV is not herein considered because only two respondents were associated exclusively with the factor, and the factor scores presented no clear image of the meaning of the attitude. Conceivably, factor IV represents a vestige of historical segregationism, for it assigned a –4 to the statement (47) that "bringing the Negro into the mainstream of American life should be a 'compelling state interest' of the highest order."

17. "Minority Firemen to Appeal Ruling against Quota Hiring," *Birmingham News*, December 28, 1978.

18. Seymour Martin Lipset and William Schneider, "The Bakke Case: How Would It Be Decided at the Bar of Public Opinion?" *Public Opinion* 1 (March/April, 1978): 42.

Appendix 7A: Q-Sample Statements and Scores of Principal Factors

Statement	Factor I	Factor II	Factor III
1. The *Bakke* decision jeopardizes the authority of the federal executive order which requires that enterprises holding federal contracts take affirmative action to correct disproportionately low employment of racial minorities.	0	−2	−4
2. *Bakke* is a manifestation of a conservative trend in American life and values, a trend which will determine, far more forcefully than any court case, the fate of affirmative action.	+3	+5	−2
3. *Bakke* will have a chilling psychological impact on many minority individuals, reducing both the hopefulness and the assertiveness with which they seek employment, promotion, special training, and the like.	+2	−1	−4
4. Isn't it strange how so many people who were never very worried about racial discrimination in the past suddenly are filled with righteous indignation because a *white* has allegedly been victimized?	+4	−1	0
5. With regard to affirmative action hiring by private business when it has been determined that minorities have been discriminated against, numerical remedies, including quotas, continue, after *Bakke,* to be permissible.	+1	+3	−1
6. The *Bakke* decision will be taken by some as an excuse for inaction or the dismantling of affirmative action programs in employment.	+4	+1	+2
7. The meaning of the *Bakke* case is clear: no more special favors. It was a victory for the individual.	−5	−1	0
8. There is no question about it—ever since the Supreme Court first decided to hear the *Bakke* case, affirmative action programs have been on "hold."	+3	−3	+1
9. The *Bakke* decision signifies that fixed quota affirmative action programs in the hiring area are in serious jeopardy, with the likelihood of termination in post-*Bakke* cases.	+2	+1	+1
10. The *Bakke* case will have very little if any impact on federal efforts to combat job discrimination against minorities.	0	−4	+5

11. The Supreme Court's decision that race is a proper factor in affirmative action means that much new progress can now be made in minority hiring and employment. 0 −4 +1

12. The effect of *Bakke* will be, if anything, to worsen the already formidable problem of nonwhite unemployment. +1 −2 −5

13. Properly designed minority race-sensitive programs can both achieve equal opportunity and prevent racial discrimination. +2 −4 +5

14. *Bakke* is a landmark case, but we don't know what it marks. We know it's terribly important, but we won't know for many years what the impact will be. +3 +3 +2

15. Special training programs and preferential promotions for minority employees in private business and industry will be among the major casualties of *Bakke,* because employers will see the decision as casting doubt upon the legality of such practices. +1 0 −3

16. The *Bakke* decision has dealt a more crippling blow to minority interests and rights than any since the decision to withdraw occupation troops from the Reconstruction South in the late-nineteenth century. −3 −5 −5

17. Supreme Court deference in principle to race-sensitive programs of selection is unconscionable, for such deference denies to whites even the limited constitutional protection afforded to blacks during the "separate but equal" era. −5 +2 −3

18. An employer should not be required to hire a less qualified person in preference to a better qualified person. −2 +5 +4

19. For a black to attain a position of responsibility, he or she must struggle harder than most whites. +5 +3 +4

20. The only criterion for colleges as well as employers should be qualification, not race, which seems to be our obsession now. −4 +4 −2

21. The tendency of *Bakke* is to strengthen the legal position of union rules and labor-management agreements that even in the past were powerful obstacles to affirmative action in hiring, seniority, and on-the-job training. +1 0 −1

22. I predict, in light of the *Bakke* decision, that the Supreme Court will soon review and deny the constitutionality of federal legislation giving preference to minority business enterprises. −2 0 −4

23. The *Bakke* decision tends to strengthen the legal position of those who insist upon affirmative action in business firms, at least when there is evidence or admission of past discrimination against minorities. −1 0 +1

Statement			
24. Decisions such as *Bakke,* condoning in principle race-conscious affirmative action programs, must have the inevitable consequence of increasing the number of mediocre or unqualified applicants for employment, professional school admission, and the like.	−3	+1	−4
25. The Supreme Court was wise in repudiating the use of racial quotas.	−3	+2	+4
26. When discrimination, particularly in employment and education, has been long and widely practiced, it cannot be satisfactorily eliminated by adoption of neutral, color blind standards for selection.	+5	−2	+1
27. Any system under which considerations of relative abilities and qualifications are subordinated to considerations of race in determining who is to be hired and promoted in order to establish a certain numerical position has the attributes of a quota system which is legally impermissible.	−4	0	+4
28. *Bakke* was a split decision. This causes anxiety, for example, in those hoping to be hired under affirmative action.	+3	+1	−2
29. *Bakke* is the most significant civil rights case since *Brown* v. *Board of Education.*	−2	+2	0
30. If, as in Powell's regard for a diverse student body, the Supreme Court seeks justifications for preferential treatment, it will find far *stronger* justifications in the employment area than it did in professional school admissions.	−1	0	+3
31. The difficulties for EEOC are compounded, because many employers, following the *Bakke* decision, will question the authority of federal laws and regulations for the protection of minorities.	+1	+1	−1
32. The *Bakke* decision was so narrow as to render it of rather little importance.	−2	−5	−3
33. Thanks to the *Bakke* case, minorities and females will have their pride restored. Whenever they are chosen for a job, promotion, or college admission they will have the confidence of knowing it is for their ability and not to fill a quota.	−2	−1	0
34. If the Constitution prohibits exclusion of blacks and other minorities on racial grounds, it cannot permit the exclusion of whites on similar grounds.	−1	+4	+5
35. Many employers will view *Bakke* as incompatible with, and superseding, an earlier court decision under which employers on employment tests might be required to accept a lower passing grade for minority applicants than for whites.	+1	0	−1

36. I believe, in light of *Bakke,* that the Supreme Court may soon overrule an earlier decision in which it allowed for racially determined seniority credit in unions and private enterprises to compensate for previous discrimination.	–1	–1	–3
37. Anyone who truly cherishes human rights must admit the danger posed to such rights by the Regents' claim (in *Bakke*) that the Fourteenth Amendment is only for the protection of specific minority groups, not for the relief of individual victims of discrimination.	–3	+3	+1
38. It would be inequitable and inconsistent for the Supreme Court, after providing relief for Bakke, to deny the same to a blue-collar white denied employment or promotion as a result of circumstances virtually identical to those pertaining to Bakke's rejection by the medical school.	0	+4	+2
39. Preferential programs only reinforce common stereotypes that certain groups are unable to achieve success without special protection.	–4	+2	+2
40. The *Bakke* decision flouts the principle of democracy, for the vast majority of Americans oppose any system of racial preference in admissions, hiring, promotion, or job training.	–1	–4	–1
41. *Bakke* will bring on more litigation because many will want to test whether the *Bakke* principles apply in private employment or in government contract programs.	+4	+4	–1
42. Large numbers of Americans see *Bakke* as *the* decision determining the legality of all preferential policies and practices, including those in the employment field.	+2	0	0
43. It was a tragic mistake in the decision written by Powell to deny the legality of racial quotas in selection processes.	+3	–3	–5
44. Judicial review is most legitimately employed to protect minority rights; *majority* rights can be effectively safeguarded by the normal *political* process.	+4	–5	0
45. The American system of constitutional liberties is undermined by any majority opinion sanctioning the use of race in the decision-making processes of public or governmental agencies.	–5	+3	–2
46. The *Bakke* decision increases the problems and difficulties for advocates and administrators of affirmative action programs for minority employment in state and local government.	+2	+2	0
47. Bringing the Negro into the mainstream of American life should be a "compelling state interest" of the highest order.	+5	+1	+3

Statement			
48. The decision supports the nation's continuing effort to live up to its historic promise—to bring minorities into the mainstream of American society.	−3	−3	+3
49. Present circumstances in the Supreme Court are volatile, despite the illusion of restraint and caution fostered by Powell's opinion. Significant change in the legal status of affirmative action is coming soon, the direction of this change to be determined by the views of the next appointees to the court.	0	+2	−2
50. The legitimate goal of bringing the disadvantaged into the mainstream of American life could be accomplished through programs which consider problems of culture, education, and income but are racially neutral or color blind.	−4	+5	+3
51. The *Bakke* decision reduces the authority of government agencies to act to prevent past and present racial discrimination from being carried into the future.	0	−3	−2
52. In recent years the gap between white and nonwhite income has narrowed. *Bakke* and like decisions will make it considerably more difficult for this trend to be continued.	−1	−3	−3
53. The *Bakke* decision will have a greater effect on the labor market than on school admissions.	0	−2	0
54. The *Bakke* decision is a gain for affirmative action in that in it the Supreme Court affirms in principle its legality. Those committed to affirmative action in employment have reason, therefore, to appreciate the implications of *Bakke.*	0	−2	+3
55. The *Bakke* decision will carry over into employment cases, especially since past discrimination within a company can often be proven.	+2	−1	+2
56. The big companies are not going to change at all as a result of *Bakke.* In affirmative action they will maintain what they know to be their social responsibility.	−2	−2	+2

State Correlates of EEOC Complaints

Michele Hoyman

The passage of the Civil Rights Act of 1964, and specifically title VII of the act, is one of the most significant labor policy developments of recent years.[1] Title VII outlaws discrimination in employment by an employer, union, or employment agency on the basis of race, sex, religion, color, creed, or national origin. The act covers employment establishments with more than fifteen employees, including public-sector employees.[2] The act prohibits discriminatory practices of all types including those involving hiring, testing, seniority, promotion, layoff, recall, and terms and conditions of employment.

The purpose here is to determine what factors are associated with the frequency of charges by states which are filed with the Equal Employment Opportunity Commission (EEOC). The EEOC charges in this study are those filed against the union or jointly against the union and the employer during fiscal years 1973-1975. While some studies of policy implementation assume that the exercise of statutory rights is automatic and they focus on how the administration of the act works, this study focuses on the structural factors which lead people to exercise their statutory rights in this policy area.

The procedures for compliance start when an individual brings a charge of discrimination to the EEOC, the enforcing agency. If investigation indicates that the charge has merit, the agency issues a complaint with an appropriate remedy. (There are some limited cases in which the agency may initiate compliance activity, specifically where there is a pattern and practice of discrimination by a party.) After the agency attempts to achieve voluntary compliance through conciliation, it may seek compliance through the federal courts. The individual who is aggrieved has a right to sue under the act. Within 180 days of filing a charge or after the agency has issued a finding of no probable cause, the charging party may obtain a right-to-sue letter. The individual can then bring a private civil suit provided he does it within ninety days of the issuance of a right-to-sue letter.[3]

It is clear that the passage of this act created a new set of obligations for both employers and unions as well as new rights for employees. Prior to this, there had been no federal guarantees of nondiscrimination specifically in employment. There had been a somewhat limited protection from discrimination under the Civil Rights Act of 1866 (42 U.S.C. Sect. 1981) which applied to race

The author wishes to acknowledge a dissertation fellowship from American Association of University Women, which made the early part of this research possible, and the efforts of Elyse Glassberg who was the research assistant.

but not to sex. However, this act was designed to protect blacks from discrimination due to state action. Even though it was interpreted to cover private actions, its application to the employment area was limited. Therefore the implication of title VII was to impose a specific employer requirement not to discriminate in personnel practices, where these practices formerly had not been regulated. This act and its interpretation by the courts has led to a myriad of changes in employment practices. Proscribed practices now range from overtly discriminatory ones, such as exclusion of women and blacks from certain jobs and use of wage differentials for blacks and whites performing the same job, to personnel practices that are neutral on their face but discriminatory in effect. The latter include job requirements such as a high school education, a clean arrest record, and a lack of child-rearing obligations where these are not job-related.

The act changed the union's obligations as well as the employer's. Unions did have slightly different obligations from employers prior to the passage of the 1964 Civil Rights Act. The union's obligations grew out of its duty as the exclusive bargaining agent to fairly represent all workers under the National Labor Relations Act.[4] The National Labor Relations Board (NLRB) was reluctant to regulate in the area of employment discrimination because the National Labor Relations Act was designed to protect employees from discrimination based on union activity.[5] However, certain union activities such as establishing separate locals by race or bargaining for differential benefits for blacks than for whites were proscribed by the courts even under the narrow test provided by the National Labor Relations Act and the Railway Labor Act.[6] Under title VII, unions have a broad mandate not to discriminate in the establishment of their requirements for membership, in the bargaining of contract language which determines employment practices, and in the treatment of their own employees.

The implication of title VII was perhaps greatest in terms of the individual employee. Not only did employees gain a statutory guarantee of nondiscrimination on the job based on their membership in protected categories but they gained a forum through which to press complaints which was independent of the employer, or in the case of unionized settings, the union and the employer. Thus the passage of the act not only implied certain potential changes in the employment conditions of those groups who were now protected from discrimination, but also implied a potential change in the power relationship which typified the traditional employment relationship. In other words, the decisions on employment procedures which were made unilaterally by the employer in nonunionized settings and bilaterally by the employer and union in unionized settings prior to the act, are now constrained somewhat by the individual employee's new status as practically a third party.

Conceptualization

What does it mean to file an EEOC charge? Filing an EEOC charge can be considered an act of protest inasmuch as the complainant is expanding the conflict beyond the normal channels of talking to a supervisor or filing a grievance. Normally, if there is a dispute between the employee and the employer, the employee attempts to solve the problem within the firm or at most extends the issue to an outside neutral, for example, an arbitrator in unionized settings where there is a collective bargaining agreement. The act of filing is what Hirschmann calls *voice.*[7] Hirschmann describes three options open to a person if he is dissatisfied with an organization: exit, voice, and loyalty. *Exit* is the decision of a dissatisfied person to leave the organization. *Voice* is the process by which a person tries to make a change but does it by staying within the organization. *Loyalty* is the option of silence and remaining in the organization. The study of why people express voice in terms of filing EEOC charges becomes important to compliance since the enforcing agency becomes involved only when a charge is filed.

The question addressed here is whether there are characteristics of the fifty states which explain why in some states people are expressing their statutory rights and in others they are not, or are not doing it at as great a rate. It is worth noting that filing complaints happens at the level of a firm, not a state. Why then is this analysis occurring at the state level? First, there are no data available based on interviews with people who have filed complaints. Neither the names of charging parties nor the firms are available since the confidentiality of the charges is guaranteed under the act. Second, the states do vary in their socioeconomic structure, political structure, and labor policy. It is important to see whether this structural variation impacts on nondiscrimination compliance in employment.

If people do not exercise their federally guaranteed statutory rights evenly across the states, this has obvious implications in terms of public policy. It means that access to this policy is unevenly distributed across the states. Access to this policy is guaranteed by people exercising their statutory rights by filing complaints. The more uneven the access to this policy across the states, as indicated by the variation in the number of complaints per thousand employees, the more the policy approaches the quality of state-by-state legislation, which means there is no nationwide minimum level of compliance effort. Conversely, the more even the distribution of access to the act is by state, the more it approaches functioning like the federal policy which it was intended to be. Thus this study addresses the limits a federalist structure places on the delivery of benefits from a national policy such as this one outlawing discrimination.

Hypotheses

Region

What key factors are expected to correlate most highly with the frequency of complaints? The region hypothesis postulates that because Southern states have had a history of institutionalized discrimination, there will be a greater need for compliance, and thus more complaint activity than in non-Southern states. This implies that the absence of a charge means there is no need for a charge to be filed. The social structure hypothesis predicts that Southern states will produce a significantly lower average number of charges than non-Southern states because the same forces in the social structure which lead to a greater need for compliance also lead to a greater reluctance to file charges. This hypothesis implies that the absence of a charge indicates nondecision rather than compliance necessarily.[8]

Socioeconomic and Political Structure

There will probably be other more important factors which vary state by state. Inasmuch as state characteristics, both socioeconomic and political, affect policy outcomes, these factors may also create an atmosphere conducive to filing complaints. It is expected that there will be a positive relationship beteen the state's socioeconomic status level and the number of complaints filed. A rival hypothesis is that it will be the state's political structure, specifically its degree of party competition, which will create a liberal atmosphere in terms of filing complaints.

These rival hypotheses are based on the debate in the public policy field as to whether socioeconomic or political structure characteristics explain more of the variation in the liberalness of state policies. The early works by Key and Lockard suggested that political structure determines the policies adopted by the state, and that the degree of party competition determines the liberalness of policy responses.[9] However, more recent works have partly refuted this conclusion, arguing that the socioeconomic characteristics of the state dominate the political characteristics as determinants of policy outcomes.[10] It has also been argued that perhaps political structure is the intervening variable between socioeconomic characteristics and policy.[11] The many divergent results may be due to the differences in operationalization of both the socioeconomic and political variables, as well as the fact that different policies were being studied in each case. Thus there is no resolution to the debate as to which are most important. It is reasonable to conclude that the relationship between political structure and policy is not as direct or as important as it seemed at first. On the other hand, the large variation in results suggests that it would be premature to discard the political structure variables, particularly as intervening variables.

The socioeconomic factors considered important to filing complaints are median income and the percentages of the population with a low education, urbanized, unemployed, nonwhite other than blacks, black, poor, and female.[12] The higher the income level and the percentages urbanized, nonwhite, female, and black, the higher the number of charges per hundred thousand in the population expected. The variables which should have a negative effect on filing complaints based on their manner of construction are percentages of low education, poor, and unemployed. The reasoning behind the unemployment relationship is: as unemployment rises, the number of complaints will decrease since employees will not want to express exit or voice due to their increased vulnerability during a time of high unemployment.

The political structure variables are (1) the vote margin for the majority party in the lower house for the elections from 1962 to 1972, (2) the degree of party competition for the governor's office from 1962 to 1972, (3) the extent of the governor's formal powers, and (4) the centralization of state legislatures.[13] In this study, the percentage of vote for the majority party in the 1972 presidential election is used as an indicator of liberalness rather than as a straight partisanship measure. This is because it was felt that the McGovern candidacy represented a more typical liberal than a typical Democratic candidate.

Political Liberalness

Northwithstanding the vast amount of literature employing the above concepts to predict policy outcomes, the feeling is that the commonly used political variables provide only general constraints within which policy is made. What is needed are some indicators of the political factors for a specific policy area. The assumption in past studies seems to have been that the same political variables will be generally important for outcomes in all policy areas. This study attempts to add some political liberalness variables which may be relevant to the area of nondiscrimination in employment. The first measure is a general one, the percentage that voted Democratic in 1972, or the vote for McGovern. There is also a measure of liberalness of the labor policy in the state, indicated by its position on the 14-B option of the Taft-Hartley Act and on the right of public employees to bargain.[14] There is also a measure of the state's commitment to a policy of nondiscrimination in employment as indicated by the passage of a fair employment practices bill which covers race or sex. The extent of commitment to public-sector bargaining can be indicated by the number of years since a bill was passed; the breadth of its coverage, specifically the degree to which the act covers groups other than police and fire; and the depth of its coverage, for example, the extent to which the law requires employers to bargain and the employer and the union to participate in certain impasse resolution devices.[15] The factors which indicate the extent of the state's nondiscrimination policy contains similar elements: (1) the age of the policy; (2) the extension of this policy to several different categories including race, color, sex, age, and marital status; and (3) the degree to which protective legislation in the state mitigates the effect of this policy of nondiscrimination.[16] (If the state protective legislation applies to both men and women, it is viewed as nondiscriminatory. If it protects only women, it is viewed as discriminatory.)

There is some discussion by Lockard and others about whether state antidiscrimination legislation does reduce actual levels of discrimination, particularly given the structure and formal powers of the state fair employment practice agencies.[17] This study does not assume a priori that the legislation has this effect, only that it may create a more liberal atmosphere.

Complaint Activity versus Compliance

There is one important qualification regarding what can be inferred from this study. The level of complaint activity may indicate the amount of administrative activity, but not necessarily the total amount of compliance. This is because compliance can occur in either a nonvoluntary or voluntary manner. Nonvoluntary compliance is compliance which results only from someone filing a successful complaint with the EEOC or their state Fair Employment Agency or a successful lawsuit. Voluntary compliance can occur in two ways: the employer (and union) may already have been in compliance before the act was passed; or the employer and the union have changed employment practices after the act was passed in order to comply, but not in response to legal action.

For the moment, assume that the worst-case analysis applies: there is no voluntary compliance occurring at all and most employers and unions are in violation of the act. In this case, the number of complaints is a fairly valid indicator of the total amount of compliance. Of course, the compliance situation probably does not meet the worst-case analysis. There probably is a certain amount of voluntary compliance in response to the law being passed if for no other reason than to avoid responding to a charge. It is beyond the scope of this study to develop independent indicators of voluntary compliance in the fifty states. Similarly, the relationship between voluntary and nonvoluntary compliance cannot be directly addressed in this study. What can be examined here is the level of procompliance activity, as measured by the number of complaints. This procompliance activity—the number of EEOC complaints—is the best available indicator of the level of nonvoluntary compliance across the states.

Although complaint activity can only directly indicate nonvoluntary compliance, it is crucial not to leave the impression that there is no link between nonvoluntary and voluntary compliance. Given the voice function that filing a complaint has, it serves to raise the issue of title VII compliance in an employment setting, whether the complaint is meritorious or not. In other words, filing a charge may serve an agenda-setting function at the employment site. A complaint may lead to filing other complaints or to voluntary efforts to achieve compliance. In this way, it is linked in a broader way to the process of compliance, both voluntary and nonvoluntary.

Results

Region does not appear to be a strong determinant of the number of complaints. Although the simple relationship between region and the number of complaints indicates that the South has a positive effect on the number of complaints ($r = .14$), there is no significant effect of region when other controls such as median income and urbanization are included.

Findings

Socioeconomic Variables

The socioeconomic variables which seem to have a large effect on the number of complaints are the median income, the degree of urbanization, and the percentage of unemployment. The simple correlation between the median income and the number of charges per hundred thousand in the population (hereinafter the rate of filing charges) is .21, and this relationship increases when other controls such as political structure and political liberalness are added. The correlation between the percentage of the population in urbanized areas and the rate of filing charges is only .07, but this frequently becomes an important negative force when other variables are controlled for. The simple correlation between the percentage of unemployed and the rate of filing charges is .49, which is in the opposite direction from what was predicted. The percentage of unemployed remains a positive and significant force in predicting the number of complaints even when other factors such as region, income level, and political liberalness are controlled for and used in a multiple regression.

Political Structure

None of the political structure variables such as the centralization of the legislature, the formal powers of the governor, the degree of party competition in the lower house, and the degree of party competition for the governor's office appear to be strongly related to the rate of filing charges, except the formal powers of the governor which has a –.34 simple correlation with the dependent variable. (However, none of these political structure variables retain much of their importance relative to the socioeconomic factors when both are included in a multiple regression.) Thus it looks as though these results support the dominance of socioeconomic characteristics over political structure in determining a state's rate of filing charges.

Political Liberalness

What about the importance of the political liberalness of the state in determining its rate of filing complaints? There were four ways that political liberalness was measured: (1) the percentage that voted Democratic in 1972; (2) the liberalness of labor policy in the state, as indicated by the 14-B provision; (3) the liberalness in terms of public-sector bargaining rights; and (4) the liberalness in terms of fair employment policy. The percentage that voted Democratic in 1972 yields some very disappointing results. (The simple correlation coefficient was -.09, which did not increase in importance when controls for region, socioeconomic structure, and political structure were included in a multiple regression with it.) Another surprising result is that the state's right-to-work policy has little effect on the rate of filing complaints.

Another indicator of the state's liberalness in labor policy is the state's policy on public-sector unionization. The only one of the public-sector bargaining variables which affects the rate of filing charges is the depth of the commitment to public-sector bargaining; that is, the extent to which the law requires the employer to bargain. However, the relationship is strong and negative instead of strong and positive. In other words, the greater the state's commitment to public-sector bargaining by guaranteeing that an employer bargain with its employees, the fewer complaints filed. A possible explanation is that in liberal states there may be less objective need to file complaints, thus resulting in a lower-filing rate. When other factors are controlled, this factor ceases to be significant.

There is only one fair employment variable which appears to have an effect on complaints once controls are added. This variable indicates the scope of the state's nondiscrimination in employment law, both the employees who are covered and the types of categories which are included. The coverage of the state law means whether public-sector employees, state contractors, unions, employment agencies, and small employers are covered as well as private-sector employers. The types of categories may include race, sex, age, and marital status, as well as protection for the principle of equal pay. This variable has a significant and negative relationship which becomes even stronger when other factors such as socioeconomic and political characteristics are added. The point is that the liberalness of the state's fair employment policy seems to have a negative rather than positive effect on the number of complaints.

Why would this be the case? There is an explanation which may be much more mechanical than theoretical. Under section 706 c of the Civil Rights Act, there is a provision requiring deferral of EEOC action for at least sixty days after proceedings have begun with a state fair employment practices agency on the same charge. (This period of time is extended to 120 days if it is the first year that such a statute is in effect.) After sixty days, if the plaintiff requests it, he has a right to file with the EEOC. The fact is that if there is a state agency, the charging party may not file with the EEOC also, at least at first. Of course, after the sixty-day period, the charging party can file with the EEOC. Therefore, what is causing this negative relationship is the ability of states with strong legislation on fair employment to handle charges effectively or at least in such a way that the person does not need to file with the EEOC. An alternative explanation is that states which are liberal enough to have comprehensive state laws of this sort might provide an atmosphere in which voluntary compliance can easily happen, thus alleviating the necessity of filing charges.

Final Model

The factors which seemed to best explain the number of charges are (1) median income, (2) percentage of urbanization, (3) the number of protected categories and extent of coverage under the state nondiscrimination law, and (4) the percentage of unemployment. Median income and the percentage of unemployment each have a strong positive effect. The percentage of urbanization and the scope of the state nondiscrimination law have a strong negative effect.

As stated before, the percentage of urbanization is not significant when it is included in a multiple regression with other variables. One statistical reason may be multicollinearity.[18] Checking for this it becomes clear that there are two variables with which the percentage of urbanization seems to be related: the median income and the nondiscrimination law's categories and coverage. The simple correlation between the median income and the percentage of urbanization is .63, while the simple correlation between the scope of the fair employment law and the percentage of urbanization is .36. Deleting one variable at a time and using the R^2 or variance explained test to determine whether there is multicollinearity, we find that there is multicollinearity between the percentage of urbanization and the scope of the fair employment variable. The fair employment variable emerges as the stronger of the two, as documented by the fact that the R^2 or the variance explained reduces to 24 percent from 38 percent when it is excluded from the regression.

Removing the nonsignificant urbanization variable from the regression reduces the variance explained in the rate of filing by only 4 percent, from 37, in table 8-1, to 33 percent as can be seen by table 8-2. Thus the regression equation in table 8-2 which includes median income, percentage of unemployment, and the scope of fair employment legislation is the better explanation of the percentage of complaints. These three variables are significant, the equation as a whole is significant, and the variance explained in the rate of filing complaints is 33 percent.

Table 8-1
Predictors of the Number of EEOC Complaints

Number of complaints per hundred thousand in labor force

= (.05) (state's median income) + (30.6) (state % unemployed)
(3.4)[a] (2.9)[a]

+ (01.2) (state's % urbanization) + (-17.7) (state's nondiscrimination law)
(-1.7) (-3.1)[a]

R^2 = .42; adjusted R^2 = .37; F = 7.9; sig = .000.

[a]Variable is significant at the .05 level.

Table 8-2
Predictors of the Number of EEOC Complaints

Number of complaints per hundred thousand in labor force

= (.03) (state's median income) + (36.1) (state's % unemployed)
(2.9)[a] (3.5)[a]

+ (-15.5) (state's nondiscrimination law)
(- 2.7)[a]

R^2 = .38; adjusted R^2 = .33; F = 9.0; sig = .000.

[a]Variable is significant at the .05 level.

Conclusions and Interpretations

Clearly, this study does not lend much support to traditional views that the more competitive the political structures, the more equalitarian the policy outcome. In fact, political structure seems to make very little difference in determining the rate of filing charges within a state. Although political structure does not seem to have much effect, political liberalness does. One of the political liberalness variables, the scope of the state's fair employment legislation, has a substantial *negative* impact on the rate of filing complaints. The direction of this relationship was predicted to be positive: the more liberal the state, the higher the number of complaints. One possible explanation is that the state agency is successfully "taking care" of a charge that would have otherwise been handled by the EEOC, under the section 706 c deferral language of the act. Another possibility is that the social structure is such that states with comprehensive coverage of employment discrimination have a conducive atmosphere for voluntary compliance, thus eliminating or reducing need for nonvoluntary compliance efforts.

This fair employment variable which has a negative effect is the only political characteristic which held strong when socioeconomic variables were added. Thus the state's liberalness on labor, as measured by the 14-B option and its policy on public sector unions, seems to matter very little.

To a great extent, socioeconomic factors seem dominant in determining compliance. Median income is positive and significant in its relationship to the rate of complaints, implying perhaps the role that information and access may play in people's expression of voice within their firm. The percentage of unemployment is positive and significant; that is, as unemployment rises, more people within a state are willing to express voice. It was predicted that both voice and exit would decrease as unemployment rose due to people's perceptions of the risk of doing either. One explanation of the positive direction between unemployment and voice may be that voice is a substitute for rather than a correlate of exit. In times of high unemployment, people perceive more risk involved in leaving the firm (exit) so they become more likely to express their dissatisfaction by filing a complaint (voice) instead.

What emerges from this study is the conclusion that socioeconomic structure may be more important than political structure in determining compliance efforts in this area. Furthermore, the political liberalness of a state, measured by either its McGovern vote or the progressiveness of its labor policy, does not make any difference in determining its number of EEOC complaints. However, the state's political liberalness in the fair employment area itself is an important factor in determining the number of complaints. The greater the state's commitment to fair employment policies, the fewer EEOC complaints. As stated in an earlier section, this can be interpreted in several different ways. However, assuming that the decisions of these state agencies are not weaker, it appears that deferral to state agencies in states with a liberal fair employment policy decreases the work load of the federal agency and achieves the same policy result.

Many factors can explain these unusual results. First, the model may not be properly specified. For example, an excluded political variable might have been more important than the socioeconomic factors. Second, there is the criticism that by choosing to aggregate the data to the state level, the important variation of complaints on the firm level was washed out. In other words, the meaning of these results may be suspect. It is true that the firm would have been an ideal unit to study, but the data are not available. If this criticism were valid, there should have been no significant relationships, just random effects. However, state-level characteristics, specifically median income, the state's employment discrimination policy, and the state's unemployment rate, are significant factors in explaining the number of complaints and together explain 33 percent of the total variation in the number of charges by state.

In terms of federalism, the results are extremely interesting. It appears that certain factors about a state's socioeconomic structure encourage the expression of federally guaranteed rights in such a way that access to this policy is not evenly distributed across the states. It is also clear that state-level policy at least in the area of fair employment practices can also make a difference. This suggests that an area for future research may be the variation in states in terms of the *quality* of the policy delivery which occurs in the equal employment area, as well as the differential access to fair employment policy.

Notes

1. The Civil Rights Act of 1964 (78 Stat. 253; 42 U.S.C. 2000e et seq.) was amended by the Equal Opportunity Act of 1972 (*Public Law,* pp. 92-261). The amendment expanded the powers of the Equal Employment Opportunity Commission to include the ability of the agency to enforce its own orders through the courts.

2. Public-sector employees and employers with fifteen to twenty-five employees were included in the 1972 amendment to the act. Before the 1964 version of the act covered only employers with more than twenty-five employees.

3. *Public Law,* pp. 82-261, section 706.

4. For a discussion of the limits of this duty, see Herbert Hill, *Black Labor and the American Legal System* (Washington, D.C.: Bureau of National Affairs, 1977), pp. 93-169; Arthur B. Smith, Jr., *Employment Discrimination Law* (Indianapolis, Ind.: Bobbs-Merrill Co., 1978), pp. 245-320; Barbara Lindemann Schlei and Paul Gross, *Employment Discrimination Law* (Washington, D.C.: Bureau of National Affairs, 1976), pp. 640-677.

5. Hill, *Black Labor,* pp. 93-94.

6. See Schlei and Gross, *Employment Discrimination Law,* pp. 640-648, for a discussion of these cases. In Steele v. Louisiana and Nashville Railroad, 323 U.S. 192, 15 LRRM 708 (1944), the Court ruled that under the Railway Labor Act the union could not bargain contract language which limits transfer opportunities for blacks; in Independent Metal Works Union Local 1 (Hughes Tools Co.), 147 NLRB 1573, 56 LRRM 1289 (1964) the Court held that the board should rescind certification for a union which had racially separate locals.

7. Albert Hirschmann, *Exit, Voice and Loyalty* (Cambridge, Mass.: Harvard University Press, 1970).

8. The region variable is based on the U.S. Bureau of the Census regional classifications of Northeast, South, North Central, and West, which were collapsed into a dichotomous variable of South and non-South.

9. V.O. Key, Jr., *American State Politics: An Introduction* (New York: Alfred A. Knopf, 1956); Duane Lockard, *New England State Politics* Princeton, N.J.: Princeton University Press, 1959).

10. Richard E. Dawson and James A. Robinson, "Inter-Party Competition, Economic Variables and Welfare Policies in the American States," *Journal of Politics* 25 (1969): 265–289; Richard I. Hofferbert, "The Relationship between Public Policy and Some Structural and Environmental Variables in the American States," *American Political Science Review* 60 (1966): 73–82; Thomas R. Dye, *Politics, Economics and the Public: Outcomes in the American States* (Chicago: Rand McNally and Company, 1966). For works which illustrate the limited effect of malapportionment, see Herbert Jacob, "The Consequences of Malapportionment: A Note of Caution," *Social Forces* 43 (1964): 256–261; Thomas R. Dye, "Malapportionment and Public Policy in the States," *Journal of Politics* 27 (1965): 586-601. See also Richard I. Hofferbert, "Ecological Development and Policy Change in the American States," *Midwest Journal of Political Science* 12 (1968): 401–418.

11. Robert L. Lineberry and Edmund P. Fowler, "Reformism and Public Policies in American Cities," *American Political Science Review* 61 (1967): 701-716; Charles Cnudde and Donald J. McCrone, "Party Competition and Welfare Policies in the American States," *American Political Science Review* 63 (1969): 858-866.

12. Data on the number of complaints filed were from the Equal Employment Opportunity Commission. The number of complaints was changed into a rate by dividing by the number in the labor force. Thus the figure is the number of complaints per hundred in the work force.

Sources for the socioeconomic variables were: median income (1970), U.S. Department of Commerce, Bureau of the Census, 1970, *General Social and Economic Characteristics,* p. 469; percentage of urbanization, Bureau of the Census, *United States Statistical Abstract,* table 141; percentage of unemployed, Census, *Statistical Abstract,* p. 469, table 141; median education, median years completed of persons 25 years and older, Census, *Statistical Abstract, 468, table 140;* percentage of low education (Under seven years education; the variable was created by adding number of females and males who completed sixth-grade education.), Census, *Statistical Abstract,* males, p. 493, table 156, females, p. 494, table 157; percentage of nonwhite, total persons of Spanish heritage plus total Asians divided by population, Census, *Statistical Abstract,* p. 470, table 142, p. 474, table 144; percentage of poor, income below 125 percent of poverty level (this is percentage of all unrelated individuals), Census, *Statistical Abstract,* p. 548, table 182; percentage of blacks, Census, *Statistical Abstract,* pp. 470–471, table 142 (this is percentage of blacks in population, not labor force); percentage of females, 16 years and over in labor force, 1970, Census, *Statistical Abstract,* p. 469, table 141; population, total and rank, Census, *Statistical Abstract,* p. 468, table 140.

13. Data sources for political structure variables are percentage of vote margin of majority party in lower house of state legislature 1964–1970, Bureau of the Census, *United States Statistical Abstract* (1973), p. 377, (1977), p. 377;

percentage of vote margin for party of governor, Census, *Statistical Abstract* (1962, 1964, 1966, 1968, 1970), p. 366, (1977), p. 505; governor's formal powers, from Joseph A. Schlesinger, "The Politics of the Executive," table 9, p. 229; Herbert Jacob and Kenneth N. Vines, *Politics in the American States: A Creative Analysis,* pp. 207-237; "Centralization of the State Legislature," in Wayne L. Francis, *Legislative Issues in the States* (Chicago: Rand McNally and Co.), pp. 74-75.

14. This provision of the Taft-Hartley Act of 1947, which amended the National Labor Relations Act of 1935, allows states to exercise the option of limiting the forms of union security which can be negotiated. The 14-B states then are those which have voted to limit union security possibilities.

15. Sources for political liberalness are: percentage of vote for McGovern (number for McGovern divided by number for Nixon plus number for McGovern), *World Almanac,* 1978 (Newspaper Enterprise Assoc., Inc., New York, N.Y.), p. 256. Source for public-sector coverage, depth of commitment, and age of the policy is U.S. Department of Labor, Labor-Management Services Administration, *Summary of Public Sector Labor Relations Policies,* 1975.

16. Sources for fair employment legislation are Bureau of National Affairs, *Labor Relations Reporter,* "Fair Employment Reporter," 8-A (1977), pp. 451:102–451:105 for categories and number of years since passage; pp. 451: 151-154 for coverage and protective legislation.

17. Although Paul Norgren and Samuel Hill, in *Toward Fair Employment* (New York: Columbia University Press, 1964), suggest that the presence of FEP laws does reduce discrimination, as indicated by the percentage of increase in nonwhite employment, Lockard refutes this. See Duane Lockard, *Toward Equal Opportunity: A Study of State and Local Anti-Discrimination Laws* (New York: Macmillan Company, 1968). Lockard maintains that the study did not include a control for the increase in the nonwhite employment and did not acknowledge the regional location of the nonprogressive states. Lockard creates his own index of improvement and his results suggest that it is not legal but social and economic factors which influence the employment gains of nonwhites.

18. Multiple regression technique assumes the independence of each variable. If two variables are related in such a way that the error terms of two variables vary together, this may cause each to lose its significance when used in a multiple regression together.

Part III
Occupational Safety and Health

9 Client Group Attitudes toward Alternative Forms of Industrial Safety Regulation

Robert A. McLean and
Ronald G. Schneck

Introduction

The control of industrial accidents has become one of if not the most widely discussed and hotly debated issues in the field of labor and employment policy in the 1970s. It has been the subject of books,[1] monographs,[2] articles,[3] and dissertations.[4] The effectiveness of existing and proposed alternative legislation has been discussed and debated in both scholarly and political circles. The nature of the debate indicates that some revision of existing legislation will be proposed within the near future. We will examine the political feasibility of possible alternative forms of legislation and the degree of support for current industrial safety legislation.

One of the chief assumptions of the research discussed here is that, when public policies have explicit client groups, it is among those client groups that political support should be measured. That is, a necessary but not sufficient condition for the acceptance of any piece of legislation is its acceptability by the group in whose interest the legislation was passed.

Industrial safety regulation clearly represents an area of public policy which has an identifiable client group: organized labor. Thus no program of industrial safety regulation is likely to be passed (or to be successful once passed) if it does not have the support of organizations representing a major part of the labor movement. The purpose of the research reported here is to determine the degree of support, at one organizational level, within the labor movement for existing and alternative forms of national industrial safety policy.

The organizational level chosen for analysis is the local union level. It is at this level that union officials have the closest contact with problems of industrial safety and health. One may expect, then, that it is at this level that effective political support for a safety policy will or will not emerge.

The authors are indebted to Sharon Noftz McLean for her aid in constructing the survey instrument employed in this study. Assistance in mailing the survey instrument was provided by the Department of Economics, University of Wisconsin-Milwaukee, and by the Milwaukee County Labor Council.

The Issues

A growing number of economists have argued that the regulation of industrial safety by government-imposed standards represents an ineffective and inefficient way to reduce the frequency of industrial accidents.[5] Standard setting is ineffective, it is argued, because the standards may not be related to the actual causes of accidents (especially the relatively large proportion of accidents caused by transitory conditions and worker practices). It is also argued that standard setting generates production inefficiency because it forces the firm to produce at a nonoptimal capital/labor ratio (compliance with standards involving the use of inefficiently large stocks of capital).

Rather, these economists argue, firms should be charged an injury tax sufficient to cover the full social cost of each accident and then allowed to find the cost-minimizing way to reduce accidents.[6] To the extent that the firm would pay the full social cost of each accident, the socially optimal level of industrial accidents would be generated by such a program.[7]

Incentive-based safety legislation has received some support from economists in the Carter Administration, though both the Secretary of Labor and the Assistant Secretary of Labor for Occupational Safety and Health have both expressed disapproval of such a program.[8] While the reluctance of the administration to adopt an incentive-based alternative to the standard-setting approach embodied in the Occupational Safety and Health Act (OSHA) of 1970 may be rooted in professional analysis, it may also be rooted in a belief that such a program would be unacceptable to organized labor.

Previous Research on Labor's Attitudes toward OSHA

While no previous research has dealt explicitly with the issue of political support for alternative forms of industrial safety regulation, there have been several efforts to catalog union attitudes toward OSHA and the general issue of industrial safety. Kochan, Lipsky, and Dyer found that 75 percent of a sample of local union activists felt industrial safety to be a very important issue in the quality of work life.[9]

Perkel and Wood, each writing from his own experience, have presented two papers on the effectiveness of OSHA.[10] Several salient features emerge from these two articles. First, each author is very dissatisfied with the administration of the current law. Frequency of inspection, delays in enforcement, low penalties, and granting variances for "economic" reasons are cited among the sources of administrative dissatisfaction. Second, the level of funding for the administration of OSHA is cited as being inadequate (reflecting, it is believed, a low commitment to the law on the part of the administrations in office at their writing). Third, especially in Wood's paper, there is a reluctance to allow market considerations to influence policy decisions in the health and safety area.

Employing a methodology similar to that used in our research, Zulusky reported on the opinions of local presidents of the Allied Industrial Workers of America toward the enforcement of industrial health and safety laws.[11] He was more concerned with objective experience than with opinions, but several conclusions as to local officers' opinions were drawn. State industrial safety programs were believed not to be effective. Further, work participation in the inspection process was believed to be an important contributor to the effectiveness of any safety regulation program.

Research Methodology

To ascertain the attitudes of local union officers toward OSHA and its incentive-based alternatives, a questionnaire was developed requesting information on the following items:

1. The industry in which the members of the officer's local were employed
2. The size of the local
3. Whether or not the officer had ever participated in an OSHA inspection
4. The age, race, and sex of the officer

A series of twelve statements was then offered, on which the respondents were asked to indicate their degree of agreement or disagreement (agree strongly, agree somewhat, neither agree nor disagree, disagree somewhat, disagree strongly). These questions dealt with the following issues: effectiveness of OSHA in reducing the frequency of occupational injuries, effectiveness of OSHA in controlling occupational illness, adequacy of resources devoted to OSHA, desirability of union participation in the administration of OSHA, the general efficacy of standard setting as a means of regulating industrial safety and health, the general efficacy of an incentive-based approach as a means of regulating industrial safety, the comparison of an incentive-based approach to a standard-setting approach, the general need for government regulation of industrial safety, and the desirability of the use of economic considerations in setting safety standards. To the response of each respondent was added the lost workday frequency rate of the first industry listed in (1) above.[12]

The survey was mailed to one officer of each affiliate of the Milwaukee County Labor Council.[13] Of the 205 questionnaires mailed, 1 was returned as undeliverable. Ninety-six responses were received (47.1 percent).[14]

The Hypotheses

Each statement with which the respondents were asked to agree was coded on a scale of 1 to 5, 5 indicating strong agreement and 1, strong disagreement. A response of 3 would therefore indicate neutrality (neither agreement nor disagreement). Thus in testable form, many of the hypotheses are stated in terms of testing for the equality of the mean response to the neutral standard of 3.

Our first task was to determine the general level of satisfaction with OSHA among the numbers of our sample. Thus we formulated hypothesis 1: the respondents are neutral as to the performance of OSHA in reducing the frequency of industrial accidents, and hypothesis 2: the respondents are neutral as to the performance of OSHA in reducing the incidence of occupational illness. It is anticipated that these two hypotheses would be rejected in favor of a negative response (mean response less than 3).

The greater the frequency of disabling injuries suffered by the members of a local union, the less satisfied should be the officers of that union with current safety regulation practices. Thus we formulated hypothesis 3: the correlation between satisfaction with OSHA and the injury frequency rate in the industry in which respondents' constituents are employed is zero. It was anticipated that this hypothesis would be rejected in favor of a negative correlation.

Casual evidence indicated that employers who have participated in OSHA inspections disapprove of the law to a lesser extent than do employers who have not participated in an OSHA inspection.[15] This difference may be due to the discovery that the inspection is not threatening to management's control of the enterprise. Thus one would expect that union officials who have participated in an OSHA inspection would register less approval of the law than those who have not participated, and who therefore still adhere to the belief that the inspection represents an effective way to force management to alter its plant arrangement. Thus we formulated hypothesis 4: the satisfaction with OSHA (both as a means of controlling accidents and as a means of controlling the incidence of occupational illness) is no different for those union officials who have participated in OSHA inspections than for those who have not. It was anticipated that this hypothesis would be rejected in favor of greater satisfaction for those who have not participated in OSHA inspections.

The literature surveyed in the above section suggests that union officers are distrustful of market or incentive-based programs toward the regulation of industrial safety. Thus we formulated hypothesis 5: respondents are neutral as to their belief in the efficacy of unregulated markets as a means of controlling industrial safety (mean response = 3). It was expected that hypothesis 5 would be rejected in favor of disagreement with the statement that unregulated markets will adequately control industrial safety. We formulated hypothesis 6: respondents are neutral as to their agreement with the statement that economic considerations should be taken into account in setting standards (mean response = 3). It was expected that, even if standard setting remains the basic approach to safety regulation, local union officials would object to the intrusion of economic considerations into the standard-setting process. Thus it was anticipated that hypothesis 6 would be rejected in favor of a disagreement with the statement that economic considerations should be taken into account in standard setting; hypothesis 7: respondents are neutral as to their approval of the general efficacy of an injury tax (as an example of an incentive-based program) in reducing the incidence of industrial accidents (mean response = 3).

It was anticipated that hypothesis 7 would be rejected in favor of the finding that respondents, on average, disagreed with the statement that an injury tax would be an effective means of controlling the incidence of industrial accidents; hypothesis 8: respondents do not believe that there would be any difference between the effectiveness of an injury tax and the effectiveness of standard setting as a means of controlling the incidence of industrial accidents. It was anticipated that hypothesis 8 would be rejected in favor of the finding that standard setting is believed to be a more effective means of controlling accidents than an injury tax would be.

While testing hypotheses 1-8 constitutes the principal task of the current research, testing several other intrinsically interesting hypotheses is made possible by the presence of information concerning respondents' characteristics on the survey instruments; hypothesis 9: there is no correlation between the size of the local union of which the respondent is an officer and the degree to which he agrees that OSHA has been effective in reducing the incidence of industrial accidents and industrial illness.

To the extent that larger organizational size generates greater separation of the experiences of officers from members, it was expected that hypothesis 9 would be rejected in favor of a positive correlation between size of local and degree of satisfaction with OSHA; hypothesis 10: there is no difference in satisfaction with the accident and illness control effectiveness of OSHA between male and female respondents; hypothesis 11: there is no difference in satisfaction with the accident and illness control effectiveness of OSHA between white and nonwhite respondents; and hypothesis 12: there is no correlation between age of respondent and degree of satisfaction with the accident and illness control effectiveness of OSHA. The authors were, a priori, unable to predict the direction of the sex, race, and age differences tested for in hypotheses 10-12.

Empirical Results

To generate a manageable number of tests, a factor analysis was performed on the raw data. The first factor (accounting for 24.5 percent of the total variance in the data matrix) clearly loaded on the responses to the twelve statements in such a way as to identify it as being related to satisfaction with current industrial safety legislation.[16] Thus for all tests in which satisfaction with OSHA is hypothesized to be unrelated to some other variable, the test was performed using both the responses to the question, "the Occupational Safety and Health Act (OSHA) has been effective in reducing the number of injuries in the workplace," and the factor scores for the first factor (herein referred to as the satisfaction factor). Thus the responses to a number of the other questions are subsumed in the satisfaction factor rather than being employed directly in testing any of the hypotheses.

The results of the tests of hypotheses 1 and 2 were quite striking. The hypotheses that respondents are neutral as to the impact of OSHA must be rejected for both the full sample and for that part of the sample representing the private sector only. What is surprising is that, counter to the expectation of the authors, the respondents expressed (somewhat limited) approval of OSHA as a deterrent to both industrial accidents and industrial disease. Previous pronouncements on these issues must then be called into question.

Contrary to the authors' expectation, hypothesis 3 cannot be rejected. The degree of satisfaction with OSHA appears to be unrelated to the injury frequency rate experienced by the members of the union officers' locals.

The result of test of hypothesis 4 indicated that the hypothesis must be rejected. The expectation of the authors as to the sign of the difference in satisfaction with OSHA was incorrect. Like management personnel, those union officers who have personally participated in an OSHA inspection exhibit a more favorable response to the law than those who have not so participated. It may be that a personal experience with the enforcement of the act instills a feeling of participation in the safety regulation process for both labor and management, thus increasing the satisfaction of the participant with the process (whatever the outcome of the inspection).

Results of the test of hypothesis 5 indicated that, as expected by the authors, respondents showed considerable consensus in their perceived need for some form of government intervention in the regulation of industrial safety. Hypotheses 5 and 6 must be rejected. The respondents were generally in agreement with Wood in their distrust of market forces to provide adequately for industrial safety.[17]

Results of the tests of hypotheses 7 and 8 showed that both hypotheses must be rejected. While respondents showed limited approval for the use of an injury tax, they appear to prefer the standard-setting approach to the use of such a tax.

Hypothesis 9 must be rejected. A significant positive correlation exists between size of membership in the officers' locals and their degree of agreement with the statement that OSHA has been effective in reducing the frequency of industrial accidents. There is, however, no significant correlation between membership size and the overall satisfaction factor. Support is therefore given to the belief that size of membership tends to generate removal of the union officer from the membership's experience.

Results of the tests of hypotheses 10, 11, and 12 were mixed. Hypothesis 10 cannot be rejected. There appear to be no sex differences in the degree of satisfaction of respondents with OSHA.[18] Hypotheses 11 and 12 must be rejected. There appear to be differences between whites and nonwhites in their degrees of satisfaction with OSHA. White respondents tend to exhibit both greater agreement with the question dealing with satisfaction with OSHA and to have significantly higher scores on the overall satisfaction factor than do nonwhite respondents. The reason for these differences is not revealed by these data, but may be due to nonwhite workers' being employed in more injury-prone jobs than white workers.

There also appears to be a significant positive correlation between age of respondent and degree of satisfaction with OSHA (however measured). The older the officers, the greater the degree of satisfaction. This correlation may be due to older union officials' having been longer removed from those jobs (usually assigned to younger workers) which are more injury-prone.

Conclusions

The conclusions to be drawn from this study are straightforward. There is some support among local union officials for the belief that the Occupational Safety and Health Act of 1970 has been effective in the control of industrial accidents and illnesses. Whites appear to support that belief to a greater extent than do nonwhites, older union officers tend to support it to a greater extent than do younger union officers, those who have participated in OSHA inspections support it to a greater extent than do those who have not participated in such inspections, and officers of large local unions support it to a greater extent than do officers of small local unions.

In terms of judging the acceptability of alternative forms of safety regulation, several findings emerge. There is a general distrust of market-based safety regulation among the union officers surveyed. While some support for the use of an injury tax is expressed, there is a strong preference among the respondents for the use of safety standards rather than reliance on an injury tax.

The respondents to the survey expressed strong support for the belief that the setting of safety standards is the best way to regulate both industrial injuries and industrial illness. Further, there was significant disagreement with the statement that economic considerations should be taken into account in the setting of safety standards.

To the extent that the support of the labor movement is necessary to the success of any reform in current industrial safety regulation, policymakers should, according to the results of this study, look elsewhere than to a purely incentive-based regulatory scheme. Those who would replace OSHA with an injury tax law, whatever may be their justification on economic grounds, are likely to find their goal rendered unattainable by political factors. The principal client group prefers to have industrial safety controlled by a rule-making rather than by an incentive-generating process.

Notes

1. Nicholas Ashford, *Crisis in the Workplace* (Cambridge, Mass.: MIT Press, 1976).

2. Robert Stewart Smith, *The Occupational Safety and Health Act* (Washington, D.C.: American Enterprise Institute, 1976).

3. Robert A. McLean, Wayne R. Wendling, and Paul R. Neergaard, "Compensating Wage Differentials for Hazardous Work: An Empirical Analysis," *Quarterly Review of Economics and Business* 18, no. 3 (Autumn 1978): 97-108.

4. Wayne R. Wendling, "Industrial Safety and Collective Bargaining: An Economic Analysis" (Ph.D. diss., University of Wisconsin-Milwaukee, 1977).

5. See especially Smith, *Occupational Safety*.

6. Robert S. Smith, "The Feasibility of an 'Injury Tax' Approach to Occupational Safety," *Law and Contemporary Problems* 38, no. 3 (Summer-Autumn 1974): 730-744.

7. The socially optimal level of industrial accidents is that level which minimizes the sum of accident costs and accident prevention costs. Guido Calabresi, *The Cost of Accidents* (New Haven, Conn.: Yale University Press, 1970), p. 26.

8. James C. Hyatt, "Proposal Labor Agency Quit Safety Role Makes Political Heat for Administration," *The Wall Street Journal*, July 18, 1977, p. 10.

9. Thomas A. Kochan, David B. Lipsky, and Lee Dyer, "Collective Bargaining and the Quality of Work: The Views of Local Union Activists," *Proceedings of the Twenty-Seventh Annual Winter Meeting of the Industrial Relations Research Association, 1974*, p. 155.

10. George Perkel, "A Labor View of the Occupational Safety and Health Act," *Labor Law Journal* 23, no. 8 (August 1972): 511-517; Michael Wood, "An Assessment of Three Years of OSHA: Labor View," *Proceedings of the Twenty-Seventh Annual Winter Meeting of the Industrial Relations Research Association, 1974*, pp. 43-51.

11. John Zalucky, "The Worker Views the Enforcement of Safety Laws," *Labor Law Journal* 26, no. 4 (April 1975): 224-235.

12. While the survey was administered in the spring of 1978, the most recent injury frequency data available for Wisconsin were for 1975. See Wisconsin Department of Industry, Labor, and Human Relations, *Wisconsin Occupational Injury and Illness Rates, 1975*, Madison, 1977. Previous research has shown the invariance in relative injury frequency rates over time. See McLean, Wendling, and Neergaard, "Compensating Wage Differentials."

13. It was the intent of the authors to mail the questionnaire to local union presidents only. The mailing list of the Milwaukee County Labor Council while including only one name per affiliated local contained some officers who held other offices than that of president.

14. Late responses continued to arrive up to six months after this writing. Three such late responses could not be included in the tests reported and are not included in the total of ninety-six.

15. "Seeing Is Believing," *The Wall Street Journal,* April 27, 1976, p. 1.

16. For a discussion of factor analysis, see John P. Van de Geer, *Introduction to Multivariate Analysis for the Social Sciences* (San Francisco: W.H. Freeman, 1971). All computations here were made using the *Statistical Package for the Social Sciences.* See Norman H. Nie, C. Hadlai Hull, Jean G. Jenkins, Karin Steinbrenner, and Dale H. Bent, *Statistical Package for the Social Sciences,* 2d ed. (New York: McGraw-Hill, 1975).

17. Wood, "Assessment."

18. Female respondents, however, seem to be less likely to believe that economic considerations should be taken into account in standard setting than are male respondents.

Part IV
Labor Law Reform

10 Toil and Trouble: Reform of the Labor Law

Charles Bulmer and
John L. Carmichael, Jr.

Introduction

The effectiveness of present procedures and practices under existing law to protect the interests and rights of working people to organize and bargain collectively with their employers has been called into question by many people, particularly those representing the interests of organized labor. Cases are cited to demonstrate that under existing law and procedure certain employers are able to defy orders of the National Labor Relations Board (NLRB) with apparently little effect. Employers are able to take punitive action against employees seeking to organize their fellow workers and the law does not adequately protect them nor does it provide for adequate compensation when workers' rights have obviously been violated. The feeling is that present law simply does not provide sufficient penalties for employers guilty of violating the law and therefore does not act as a deterrent to such actions on the part of the employers.

The present law also allows for long delays in holding representation elections and does not provide for quick disposition of cases in which unfair labor practices have been alleged. Arguably, the NLRB's procedures need to be changed so that employers are encouraged to respect the rights of workers and not deliberately obstruct the process of union organization.[1] The present law, it is alleged, actually provides incentives to employers to obstruct the process of union organization.

However, many employers view the proposed labor law reform as evidence of a concern on the part of labor unions that its membership has declined as a percentage of the total working force. Employers view the proposed changes in the labor law as essentially an attempt to alter this trend. Few issues in recent years have generated so much controversy as proposed labor law revision.

Expansion of Board Membership

Included in the Labor Reform Act of 1977,[2] which passed the House of Representatives but failed in the Senate due to a filibuster, was a provision expanding the total number of members of the NLRB from five to seven. Proponents argue that such increase would enable the board to expedite decisions on the growing number of cases reaching the board. Under section 3(b) of the existing law the NLRB can "delegate to any group of three or more members any or all of the powers which it may itself exercise." With the expansion of the board it can operate in panels of three, thus providing a division of labor for the board.[3]

However, critics of the proposed change assert that the backlog of cases really occurs in the work of the administrative law judges and other staff who initially handle the cases and not at the board level.[4] A related proposal for reform which would alleviate congestion of cases at this level is one which would permit decisions of administrative law judges to be final unless the board accepts a case for review. Administrative law judges are lawyers and should be well qualified to dispense with most of the cases which come to the agency.

A major criticism of the proposed expansion of board membership has been made by business opponents who insist that this increase of membership during a Democratic administration would move the board to a more prolabor point of view and not really alleviate the problem of increased caseload.[5] It is suggested that increasing membership from five to seven may produce additional delays because agreement among seven members will be required rather than the present five and this would mean even additional time for debate and discussion. However, the power of the board to operate in panels of three, and the fact that it will probably do so except in the most controversial cases, suggest that expanding the size of the board would facilitate handling cases.

Accelerating Court Appeals

Under the present law, section 10(f) provides for filing an appeal from an order of the board to the Federal Court of Appeals in the circuit wherein an alleged unfair labor practice occurred. No time limit is set for filing the appeal. In the Labor Reform Act of 1977 this section would have been amended to provide for an appeal within thirty days from a final order of the board; otherwise, the decision of the board would be final.[6]

The intent of this amendment would be to expedite a final decision on the matter by preventing extended periods of delay since theoretically the aggrieved party, at present, has an indefinite time in which to file the appeal. However, it is possible that by forcing a more timely filing of appeals the outcome will be an increase in appeals and an actual increase of cases in the federal court. Once the appeal has been filed, the court would have to rule and the time in which the final ruling takes place typically involves an extended period. If the number of appeals increases, the time period for a final ruling by the court should increase as the caseload of the court becomes larger. Furthermore, under the present law, voluntary compliance with the board's decision frequently takes place; speeding up the appeals process might obstruct the ability to secure such compliance, though there would still be the possibility of out-of-court settlements.

Additional Cases Receiving Priority Treatment by the Board

Proponents of labor law reform have advocated the addition of new categories of cases to receive priority handling by the board. Section 10(L) of the present act provides for priority treatment of cases involving certain alleged unfair labor practices, but does not include those in which an employee has been fired. Section 10(L) would be amended to include cases in which an employee has been fired during an election to determine union representation or after a union has been designated to represent the employees but before a collective bargaining agreement has been negotiated. Section 10(L), as amended, would require that the board give precedence in the disposition of these cases as well as the others contained in the section.[7]

However, there is no stipulation setting a time limit for decisions on these cases, and it is questionable whether the proposed requirement would materially affect the rapidity with which a decision will be made. Unquestionably, an employee who has been illegally fired should be reinstated without undue delay and the proposed amendment to the law would increase the likelihood that this would occur.

The board is required to seek an injunction in the Federal District Court to reinstate an employee illegally fired.[8] Additionally, the employer would have to compensate the worker for lost wages by paying him twice the wages lost during the period of unemployment. These changes would strengthen the rights of workers and conceivably benefit employers in that decisions by the board would be expedited and the amount of compensatory wages to be paid would be reduced.

Expediting Procedures and Representation Elections

In recent years a number of criticisms have been leveled at the procedures of the NLRB and in particular the process involved in setting the dates and certifying the results of representation elections. The belief is that these procedures have led to unwarranted delays in the calling of elections and that the overall process as administered by the NLRB needs to be expedited. To achieve this result the present law should be changed to speed up the process. One way this might be accomplished would be to provide for a quorum of two members of the board to decide to take the case on appeal. The change would expedite the process by limiting the necessity for review by the board and enhance the position of the administrative law judges by making their decisions final in most cases. An additional suggestion is that Congress should increase the number of administrative law judges thus reducing the caseload and blacklog of cases, further speeding up the process.

Make-Whole Clause

One of the major concerns of unions is to assure that employers will not, because of refusal to negotiate, avoid wage increases obtained industrywide. The Labor Reform Act included provisions which would impose wage settlements on employers who would not negotiate in good faith with duly certified employee representatives.[9] The so-called make-whole provisions would impose wage rates corresponding to the industrywide average. This suggestion would provide a strong incentive to employers to negotiate in good faith.

Compensation Provision

A problem which has developed under the present law relates to the issue of compensation for an employee who is illegally fired for engaging in organizing activity. Obviously, the intent of such a provision is to discourage employers from dismissing employees engaged in organizing activities by making it expensive or unprofitable for an employer to take such punitive action. The logic is, of course, if an employer has to pay an employee for the period of his illegal dismissal, the employer will not be inclined to take such action. However, the problem is that the effect of the law has not always produced the desired result. Under present law, employers—if they are guilty of illegally firing an employee—are only required to compensate the employee for the amount of his wages if he had not been illegally fired. In addition, any amount of wages which the employee received during the period of his illegal dismissal would be deducted from the amount the employer would otherwise be required to pay. Also an employee illegally dismissed, is expected under present law to seek employment elsewhere during the time of his illegal dismissal. Employers who have illegally dismissed employees for engaging in organizing activity, can even object that the employee whose rights under the law had been violated did not diligently seek comparable employment or exercised poor judgment in seeking other employment during the period of his illegal firing. The compensation for the illegal action taken by the employer can thus be reduced by the fact that the harassed employee did not show good faith in seeking other employement during the period of his harassment. The result is that in some situations there is very little penalty in the present law to discourage employers from violating the rights of employees. A suggested change in the law would provide that the compensation due to any employee who is illegally dismissed "shall be double the employee's wage rate at the time of the unfair labor practice."[10] The idea of course is that such penalties would be more likely to discourage unfair labor practices on the part of employers.

Withholding Government Contracts

One suggestion which has been advanced to pressure employers to comply with final orders of the board is to deny government contracts to employers who refuse to comply with the board's orders.[11] Government purchases account for a substantial share of many firms' total sales and withholding government contracts could have a significant impact on the affected businesses' profit margins. The pressure which would result from such lost sales would probably force employers to comply quickly with the board's orders. However, questions can be raised as to whether the government's procurement process should be used in this way. It is possible that serious constitutional problems might be raised by such a procedure. This remedy can be viewed as an improper use of government power. Secondary boycotts on the part of unions are prohibited since they enlarge the scope of the labor dispute and put too much pressure on the employer. Why, then, should the government be required to boycott firms by refusing to make purchases?

In addition, there is some question as to whether the proposal would work as well as some of its supporters believe. Many firms inclined to resist compliance with the board's orders might be inclined to resist this kind of procurement pressure as well. Some might be in a position to forgo government contracts and accept reduced business. The result could be that workers would be intimidated by the fear of loss of government contracts which could result in the loss of jobs. Workers would be faced with a loss of jobs and the proposed change would fail in its purpose to increase protection to employees.

Equal Access

The feeling is that under present procedures employers are able to campaign during working hours against union membership thereby giving the employer an unfair advantage over those attempting to organize the workers. The suggested revisions in the law would require that union organizers receive an equal opportunity to meet with workers during working hours to counter the employer's antiunion efforts.[12] The intended result would be to equalize the access to workers by both employers and union organizers. However, the law provides that organizing efforts be pursued consistent with the maintenance of normal and orderly production.

Employers assert that equal access will give labor unions an unfair advantage. Organizing efforts can now take place after hours, but to open up the working premises for organizing efforts would give the unions an unfair opportunity since unions already have an advantage over employers in meeting with workers when the business is not operating. Employers, under the proposed change, would be reluctant to express their views to workers during normal operating hours. Also it would be difficult if not impossible to maintain orderly production if solicitation of workers should occur during the workday.

Conclusion

Useful labor reform should be based on one major goal, that of providing and maintaining equality of bargaining power between employers and employees. Many of the proposed changes in the existing labor law would probably enhance the achievement of this goal. Certainly, more responsible actions on the part of employers and labor unions is desired. Although the labor law reform revision bill was defeated in 1978, attempts will probably be made in the future to revive at least some of the proposals.

The present labor law, in many instances, does not seem to be adequate to the task of protecting the rights of workers to organize and bargain collectively with their employers. Employers have devised a number of techniques which have enabled them to flout the intent if not the letter of the law. Some particularly incorrigible employers have even gone so far as to defy the board, the courts, and the law because the present legislation allows them to do so with very little or no penalty.

A system with all the inadequacies the forgoing discussion has alluded to is obviously not able to protect the legitimate interests and rights of American workers. Substantial majorities of both houses of Congress recognize that the present law has become a sham which actually works to the interest of unscrupulous employers rather than protecting the rights of American laborers. A substantial effort by certain business interests intent on maintaining the present system along with a determined filibuster by some of the more conservative elements within the Senate defeated the much needed reforms provided for in the Labor Reform Act of 1977. However, the abuses which continue under the present law will no doubt maintain labor law reform as a major policy issue for the 1980s.

Notes

1. See case of Monroe Auto Equipment Company, Hartwell, Georgia, in which a representation election had taken at least thirteen years and still had not been resolved. This is reported in Matt Witt, "Why Labor Laws Don't Work," *Juris Doctor* 7 (October 1977): 200.

2. For additional discussion of the Labor Reform Act of 1977, see Charles Bulmer and John L. Carmichael, Jr., "Revamping the Labor Laws: Some Major Proposals of the Labor Reform Act of 1977," *New Labor Review* (Fall 1978), upon which this article has drawn.

3. See proposed amendment to section 3 of the National Labor Relations Act, contained in the Labor Reform Act of 1977, *Congressional Record,* 95th Cong., 1st sess., July 19, 1977, p. 7389.

4. Discussed in *The Wall Street Journal,* 19 April, 1978, p. 24.

5. Ibid.

6. *Congressional Record,* 95th Cong., 1st sess., July 19, 1977, p. 7390.

7. Ibid., p. 7391.

8. Sect. 10(L), National Labor Relations Act (1935).

9. *Congressional Record,* 95th Cong., 1st sess., July 19, 1977, p. 7390.

10. Ibid.

11. The Labor Reform Act amends sect. 10(c) of the present act to read, "notwithstanding any other law, no contracts shall be awarded to a (person violating a final order of the board) during the three-year period immediately following the date of the (Secretary of Labor's certification to that effect)." *Congressional Record,* 95th Cong., 1st sess., July 19, 1977, p. 7390.

12. Sect. 6 of the present act would have been amended to include this provision.

11 Labor's Eroding Position in American Electoral Politics: The Impact of the Federal Election Campaign Act of 1971

Edwin M. Epstein

Although it remains a very important factor in the American political process at the outset of the 1980s, organized labor's position in American electoral politics has eroded significantly during the past decade, particularly relative to that of business. This erosion in labor's role is not a result of astute political behavior on the part of the business community. To the contrary, it is an unintended consequence of labor pursuing its own short-run political objectives. Although the Federal Election Campaign Act (FECA) of 1971, as amended in 1974 and 1976, has legitimated and facilitated the establishment of political action committees (PACs) by both business and labor, ironically it has benefited business-related groups more than unions. The ironic dimension of this development is that, during all three legislative rounds, it was organized labor that was instrumental in drafting and securing passage of the key provisions relating to PACs in order to consolidate and improve its electoral position. The federal regulatory framework which evolved during the 1970s explicitly authorizes certain labor and business electoral practices which, if not patently illegal hitherto, were under a cloud of legal uncertainty. Since the FECA restricts greatly the sums of money which individuals and political parties may contribute to federal candidates, the legislation and derivative regulations have increased substantially the importance of corporations, other business-related groups (as well as ideological and single-issue organizations and other collectivized social interests), and to a lesser extent labor unions as funding sources in federal elections. These regulatory developments, moreover, have institutionalized and legitimated the PAC as the primary vehicle for business and labor participation in electoral politics. Absent major changes in the federal regulatory framework governing elections, such as total public financing of all House and Senate races or a fundamental reversal of present congressional policy concerning political action committees, PACs affiliated with business are very likely to expand in numbers and strength during

---

This work was prepared unde a grant from the Russell Sage Foundation, New York, N.Y., administered by the Institute of Governmental Studies, University of California, Berkeley. Valuable support services were rendered by the Institute of Industrial Relations, University of California, Berkeley.

the next several years and to play an increasingly important role in federal elections. Consequently, the approximate electoral balance between business and labor which exists currently will unquestionably tip in favor of business, particularly in the area of raising and contributing monies to federal candidates through the use of PACs. Labor, of course, has not been oblivious to its declining state and is presently seeking on both the legislative[1] and judicial[2] fronts to counteract the gains made by business during the past several years both by seeking to lessen the size and importance of PAC moneys in elections and by challenging the solicitation and contributions practices of the largest corporate committees.

A final irony of these developments is that the legislative changes of the 1970s occurred within a context of political reform intended to remove or at least reduce the impact of wealthy persons and other special-interest groups on the election of federal officeholders and to enhance the influence of the average, unaffiliated citizen in the electoral process. As with all social phenomena, however, some consequences of comprehensive federal regulation of elections were sought and intended by proponents of electoral reform, while others were undesired, unanticipated, and indeed opposed by both citizen reformers and their legislative allies.[3]

This chapter will examine the consequences of federal regulation of electoral politics during the 1970s for labor unions, corporations, and other business-related groups, and will venture some thoughts regarding the implications of these developments for the American political process.[4]

Pre-1971 Federal Regulations of Electoral Activity

As James Madison recognized nearly 200 years ago in *Federalist No. 10,* the efforts of economic interests to influence government have been a staple of American politics. In a contemporary context, political involvement by labor and business has been an inevitable and enduring concomitant both of America's political heritage of democratic pluralism and of the vital importance of governmental policies and decisions to every business firm and labor union in an interdependent political economy. By means of the electoral process, business and labor organizations (in common with other social interests) have sought to influence the election of officials sympathetic to their positions and to bring before the public ballot measures supportive of their organizational needs. For unions, corporations, and other business-related groups, political action, in short, constitutes an organizational effort to respond to and shape a volatile, uncertain, and frequently hostile environment.

Electoral involvement, first by corporations around the turn of the century and, a generation later, by labor unions evoked not surprisingly, a strong reaction among those elements of American society which felt themselves to be severely disadvantaged politically by the ability of these organizations to utilize politically levels of financial and organizational resources not available to ordinary citizens. Since 1907 and 1943, respectively, corporations and labor unions have been forbidden from using their treasury funds for contributions in connection with federal elections. The War Labor Disputes Act of 1943 (commonly known as the Smith-Connally Anti-Strike Act)[5] applied to labor unions for the duration of World War II the then-existing prohibitions against corporate campaign contributions. The catalyst for this legislation was the hostility which organized labor's substantially increased political vigor and effectiveness during the 1930s had aroused among congressional conservatives. This wartime ban on union electoral activities was extended to peacetime by the provisions of the Taft-Hartley Act of 1947.[6] Thereafter, union and corporate electoral activities were regulated uniformly. Between 1947 and 1972, the basic provision governing corporate and labor electoral activity (18 U.S.C. 610) prohibited both company and union contributions and expenditures in federal primaries, and general elections and nominating conventions.[7] Two policy reasons underlay this prohibition. First was the perceived need to prevent large economic interests from dominating the selection of public officials in such a way that the integrity of the political process and of officials chosen by it would be subverted in either appearance or reality. The second was the desire to protect corporate shareholders and union members from having monies invested or contributed by them used by the management or union leadership to finance candidates and causes to which they had not assented.[8]

Union experience with PACs dates to the mid-1930s, when John L. Lewis established Labor's Non-Partisan Political League. Then in 1955, the merger of th AFL and CIO brought with it the creation of the Committee on Political Education (COPE), the model for virtually all future PACs. From the outset, national, state, and local units of COPE have not only raised and distributed funds but have also served as the mechanism for coherent and comprehensive union electoral activity, including voter registration and get-out-the-vote drives, political research on candidates and issues, polling operations, political education, phone-banking, and the rating and endorsement of candidates. Accordingly, by the time FECA went into effect on April 7, 1972, organized labor had over thirty years of experience with the political action committee.[9]

Business, however, did little with PACs before 1972. At no time during this era did the number of business-related committees much exceed fifty. The great majority of the PACs were industry- rather than company-based. Indeed, until the reforms in the campaign financing laws of the 1970s, with their strict limitations on individual donations and effective public disclosure of the sources of funds, there was little need for business PACs. Monies from business-related sources could legally enter the electoral arena, largely undetected, in virtually unlimited amounts in the form of individual contributions by wealthy persons affiliated with corporations and other business organizations. For example, the

Business Industry Political Action Committee (BIPAC), formed by affiliates of the National Association of Manufacturers during the early 1960s and the prototype business-related PAC, was a pale shadow of COPE. Prior to 1972, PACs were accordingly a minor factor in channeling business-related funds into the electoral arena. During this era, moreover, there existed an important distinction between labor and business electoral activities; a distinction which, though still pertinent today, is rapidly becoming less viable. From the onset, organized labor based its electoral strategy on the mobilization of mass political participation by union members and their families. Campaign contributions to candidates and parties, while surely important, were but one facet of union activity. Conversely, business' electoral efforts focused almost exclusively upon stimulating financial contributions activity among elites, namely, senior corporate and trade associations' executives and directors, and major shareholders, a group best reached through quiet, informal, direct solicitation efforts.[10]

The FECA of 1971 and Amendments

The Federal Election Campaign Act of 1971 and amendments, which have catalyzed the growth of business electoral activities during the 1970s, particularly the tremendous expansion in the number and size of PACs, are the root cause of American labor's eroding electoral position vis-à-vis business.[11] The 1971 act allowed corporations and labor unions to (1) communicate on any subject (including partisan politics) with stockholders and members, respectively, and their families; (2) conduct nonpartisan registration and get-out-the-vote drives directed at these same constituencies; and (3) spend company and union funds to establish and administer a "separate segregated fund" to be used for political purposes–that is, to set up PACs.[12]

The provision authorizing PACs was added to the bill on the House floor through an amendment drafted by the AFL-CIO. In this amendment, organized labor was seeking insurance against the possibility that the Supreme Court, in a case pending before it, would uphold a Court of Appeal's ruling that a PAC organized by a Pipefitters local in St. Louis was compulsory and union-financed rather than voluntary and member-financed, and was therefore illegal.[13] Uncertain as to what direction the Supreme Court would take in *Pipefitters,* the AFL-CIO sought legislative legitimization of the key aspects of its electoral efforts since the 1940s: utilization of PACs to raise and distribute monies, political communications with its members,[14] and member-oriented registration and get-out-the-vote activities.

Unions were, according to the AFL-CIO officials who helped draft the amendment authorizing PACs, taking a calculated risk. Political necessities required that in any new legislation corporations would receive rights identical to those accorded unions. Since previous corporate electoral activity had been aimed at upper-level management rather than at shareholders and had focused primarily on fund-raising activities rather than on nonfinancial endeavors, few labor leaders thought business firms would establish PACs. Union political strategists estimated, therefore, that the benefits from removing the threat to union PACs posed by the *Pipefitters* case would exceed the risks of giving business a virtual carte blanche to create PACs. Though the new 1971 law provided the basis for the Supreme Court's reversal of the Court of Appeals in *Pipefitters* (1972),[15] it nonetheless turned out to be a strategic (if perhaps unavoidable) error for labor in the longer run. Business, ironically, played no substantive role in shaping the legislation.

Corporate PACs were relatively inactive in the 1972 election. Some ninety corporate PACs operated in 1972, nearly eighty of which were established after the 1971 act went into effect. In addition to the soliciting entrepreneurship of the Finance Committee to Reelect the President (which raised substantial sums from business sources), there was another reason for the minor role of corporate PACs in the 1972 elections. Many companies with government contracts were fearful of establishing PACs after Common Cause, in a lawsuit against TRW Inc., questioned whether the section 610 authorization of corporate PACs was compatible with another section in the 1971 act (18 U.S.C. section 611, 1972) that prohibited campaign contributions by government contractors.[16]

The post-Watergate disclosures provided new impetus to campaign reform. The 1974 FECA was intended to alter fundamentally the character of campaign financing by reducing the influence of large individual and special-interest contributions, providing for public financing of presidential races, strengthening disclosure requirements, and creating the Federal Election Commission to administer federal election laws.[17] Since a number of unions were government contractors by reason of their federal manpower training and development contracts, labor had also been concerned about the Common Cause suit. Thus in the debate of the 1974 FECA amendments, union lobbyists led the successful campaign for a statutory provision clarifying that corporations and labor unions with government contracts were not prohibited from establishing PACs.[18] But as in 1971, the effort backfired—in this instance because the overwhelming majority of government contractors are corporations. Once again business, not labor, became the major beneficiary of labor's effort to secure its electoral position. With the benefit of 20-20 hindsight, union officials today consider their successful efforts to amend section 611 to have been a major strategic mistake given the fact that the clarification of section 611 has proven to be an important link in the chain of legislative developments which have resulted in the dramatic growth of corporate PACs.

SUN-PAC and the 1971 FECA

In late 1975, the newly created Federal Election Commission (FEC) issued its ruling in SUN-PAC,[19] easily the most momentous advisory opinion in its five-year history. In a controversial 4 to 2 decision, FEC held that Sun Oil Company could (1) use general treasury funds to establish, administer, and solicit contributions to SUN-PAC, its political action committee, (2) solicit contributions to SUN-PAC from both stockholders and employees, and (3) establish multiple PACs, each with separate contribution and expenditure limits, as long as the monies came solely from voluntary contributions. The FEC dissenters contended that the legislative history of the 1971 FECA and 1974 amendments indicated clearly that corporations were restricted in soliciting PAC contributions from stockholders and their families. To hold otherwise, they argued, would destroy the political balance which Congress had sought to establish between corporations and unions since a company PAC would have many more potential contributors than a union PAC which could solicit only its members and their families. (To illustrate, Sun Oil had 126,555 shareholders and 27,707 employees of whom only a small percentage was unionized. The corporation could therefore solicit over 150,000 individuals, many times more than the union.) While it was the 1971 FECA and 1974 amendments that provided the legal authority for corporate PACs, it was SUN-PAC that proved to be the energizing force for their explosion in size and numbers. In the six months following the FEC's decision, over 150 corporations established PACs, bringing the number in existence to nearly 300.

Not surprisingly, labor groups vigorously denounced the SUN-PAC ruling and resorted once again to Congress for redress. In early 1976 the Supreme Court, in *Buckley* v. *Valeo,*[20] declared a number of key provisions of the 1971 FECA as amended in 1974 to be unconstitutional, thereby requiring Congress to draft new statutory provisions. The hub of the compromise worked out by Congress within the 1976 FECA,[21] was to restrict corporate PACs to soliciting contributions from stockholders and executive or administrative personnel and their families, while, as before, labor unions were limited to raising money from union members and their families. Twice a year, however, union and corporate PACs could make use of crossover rights, that is, solicit the other's constituency by mail, using an independent third-party conduit. Organized labor achieved a key objective when it was permitted to use payroll-deduction plans (checkoffs) to collect from its members if the company PAC used that method with its stockholders or executive/administrative personnel. Finally, a "nonproliferation" provision was included: while a corporation or union could set up an unlimited number of PACs, all such affiliated committees were restricted to a contribution limit of $5,000 per candidate per election. This provision was designed to eliminate the establishment of multiple PACs by corporations seeking to take advantage of the SUN-PAC ruling. Moreover, membership organizations, trade associations, cooperatives, and corporations without capital stock (the great majority of which are clearly business-related) were explicitly authorized in the 1976 act to establish PACs.

In summary, while the 1976 amendments restored part of what organized labor had lost in SUN-PAC, they gave the business community far greater running room in the electoral process than heretofore. Ironically, as a consequence of three rounds of election legislation during the 1970s, labor and especially business-related groups are in a position to exert a greater impact upon federal electoral politics than they were at the beginning of the decade, a development which is quite contrary to the intentions of reformers who sought to free the electoral process from undue influence by "special interests."

The PAC Phenomenon

Thanks to the reporting provisions of FECA, FEC statistics for the 1975–1976 and 1977–1978 election biennia graphically reveal the rapid expansion in number and level of activity by PACs in general, and business-related PACs in particular. By using the available FEC data, it is possible, moreover, to develop an increasingly accurate picture of the demography, characteristics, and electoral behavior of PACs over the past five years. Accordingly, little disagreement exists among observers of campaign financing regarding the figures associated with the PAC phenomenon. Disagreement abounds, however, among these same observers concerning the meaning of these figures, and their implications for the electoral process. Let us examine the essential data. From the beginning of 1975 to the end of 1978, the number of PACs rose from 608 to 1,633 (1,938 PACs were registered at one point or another during the 1977–1978 campaign cycle) and total spending by these committees rose from an estimated $36.9 million in 1975–1976 to $77.8 million in 1977–1978.[22] The most recent available FEC data indicate that as of December 31, 1979, 2,000 committees were registered.[23] In 1977–1978 PACs contributed $35.1 million to congressional candidates, representing 18 percent of the $199 million the candidates received from all sources, while in 1975–1976 nonparty committees contributed an estimated $20.5 million, some 20 percent of the nearly $104.8 million received by congressional candidates.[24] By way of comparison, Common Cause estimates that in campaign 1974, interest-group donations to congressional candidates totaled $12.5 million.[25]

Focusing specifically on labor and corporate PACs, we find that until the time of FEC's SUN-PAC decision (November 1975), there were more labor than corporate PACs in existence (226 to 139). Since then, however, the preponderance has swung heavily to the corporate side (949 to 240) as of December 31, 1979. While labor PACs outraised, outspent, and outcontributed their corporate counterparts in both campaign 1976 and campaign 1978, the labor margins for each category were cut drastically for the latter campaign. In 1976 labor PACs outraised corporate PACs by $11.8 million, $18.6 million to $6.8 million. By

1978 the margin had been reduced to $2.1 million, $19.8 million to $17.7 million. In 1976 labor committees gave $8.2 million to congressional candidates compared to $4.3 million by corporate PACs; by 1978 the margin was reduced to $400,000, $10.2 million to $9.8 million.[26]

We have seen that the dramatic increase in total business PAC activity between 1974 and 1978, particularly since 1976, reflects largely the sharp rise in the number of corporate PACs. While labor PACs increased some 10 percent and those affiliated with "other interests" more than doubled their number, the number of corporate PACs has expanded almost tenfold. Let us look more closely at the demography of these PACs, particularly those of the corporate variety.

According to FEC records, there were 821 corporate PACs which operated during the 1978 election. Using the most recent *Fortune* compilations of the top 1,000 industrials and 300 leading nonindustrials (1,300 firms in all), we find 244 (24.4 percent) of the former and 124 (41.3 percent) of the latter, 368 (28 percent) in all, formed PACs. Those companies with PACs were distributed in the following fashion: 202 were affiliated with the top 500 industrials; 42 with the second 500; 124 with the leading nonindustrials (a total of 368 for *Fortune*-ranked companies) and the remaining 453 were associated with companies not ranked by *Fortune.* Thus contrary to popular belief that the corporate PAC phenomenon reflects a wholesale adoption of the PAC mechanism by American big business, 55 percent of all committees have been formed by non-*Fortune*-ranked firms (substantial though these enterprises might actually be) and 72 percent of the *Fortune*-listed companies had not established a PAC. Indeed, impressive as the growth in corporate PACs has been, what perhaps is even more significant is how few there are given the total population of potential corporate PACs. The 821 corporate PACs active in the 1978 elections (assuming for the moment they all fell into the $100 million or more category) represented only 22 percent of the 3,755 U.S. corporations with reported assets of $100 million or more (1974) and a meager 3.4 percent of the 23,834 corporations with reported assets of $10 million or more.[27] In short, the market for potential PAC formations is virtually untapped, even if we consider only the very largest business firms.

Focusing exclusively on the *Fortune*-ranked companies, there is a direct correlation between size of the company and its propensity to form a PAC. Whereas 149 (60 percent) of the top 250 industrials (including 70 percent of the leading 100) formed PACs, only 53 (21 percent) of the next 250 and 42 (8 percent) of the second 500 firms followed suit. A similar tendency is apparent among nonindustrial corporations. The higher its position on the *Fortune* list of the top fifty companies, the more likely the firm is to have a PAC. Of the 14 corporations listed by the FEC as having the top 10 PACs in terms of adjusted receipts, adjusted disbursements, and contributions to candidates, 13 were ranked by *Fortune:* 6 were in the top 50 industrials, 2 in the second 50, and 3 in the second 100, 1 in the 200 to 250 groups, and 1 was the eleventh-ranked retailer.[28] Only one firm, American Family Corporation, an insurance and financial services company, did not appear in any of the *Fortune* lists. On the basis of the available evidence, it appears to be safe to assert that in the nation's largest industrials, company size is correlated positively with both the existence of a PAC and the level of activity which the PAC maintains. A similar pattern holds true for nonindustrial firms, though there is considerable variance among industry categories. PACs associated with the largest firms were most dominant in terms of their activity among transportation companies, utilities, and particularly retailers. They were somewhat less pervasive, though still of considerable importance, among commercial banks and diversified financial companies. The number of PACs among life insurance companies is so small (only five of the top fifty firms created PACs) as to be of little analytical interest.

The higher incidence of PAC formation among *Fortune*-listed nonindustrials when compared with industrial companies is not surprising. Since only 50 firms are ranked in each category, the 300 nonindustrials tend as a cohort to be larger than their industrial counterparts. Another factor, however, is probably operative in the case of life insurance which provides a partial explanation of the tendency for *Fortune*-ranked nonindustrials to form federal PACs more often than industrial companies. This is the greater importance of the decisions of the federal government in contrast with those of state governments, to the operations of the firm. There seems to be a relationship between the importance of federal governmental decisions to the operations of an industry (or a firm) and the tendency to form PACs. In the nonindustrial category, commercial banks, diversified financial companies (a substantial number of which have savings and loan associations, consumer finance companies, mortgage banks, and security brokerages as important parts of their operations), utilities and transportation companies are all highly regulated by the federal government, or otherwise greatly affected by governmental decisions which pertain explicitly to their industry. Retailing companies, which might be considered an anomaly in this group since historically they have not been considered a part of a federally regulated industry, have been primary targets of the federal consumer legislation which has developed in the past decade and a half. As a group, the various categories of nonindustrials ranked by *Fortune* are arguably more keenly influenced by industry-specific federal policies than many industrial firms. When combined, the leading nonindustrials' relatively greater firm size as a cohort, and the importance to them of decisions from Washington help explain their increased propensity to form PACs when compared with *Fortune* industrials.

Simple corporate-labor comparisons, however, understate significantly the extent of business' electoral role during both 1976 and 1978. This is so because the FEC's classification scheme puts groups that are not explicitly corporate or labor into four separate categories: no-connected organizations, trade/membership/health, cooperatives, and corporations without stock. In 1978 the four categories accounted for 836 PACs with receipts of $43.1 million, disbursements of $43.6 million, and direct contributions to candidates of $15 million.[29] If we assume that only half the amounts raised, spent, and contributed to congressional candidates by the noncorporate, nonlabor PACs emanate from business-related committees—a very conservative estimate indeed—the receipts and disbursements attributable to business rise by over $21.5 million and contributions to congressional candidates by $7.5 million. Thus the estimated totals for aggregate corporate and business-related PAC activity for 1977–1978 are receipts, $39.3 million; disbursements, $37.1 million; and contributions to congressional candidates, $17.3 million. Based on these estimates, business and business-related groups outraised and outdisbursed labor groups by almost two to one in 1978 and outcontributed them by almost 70 percent. The largest noncorporate business-related PACs, such as those of the National Association of Realtors (receipts, $1.85 million; disbursements, $1.81 million; contributions, $1.12 million) and the National Automobile Dealers Association (receipts, $1.46 million; disbursements, $1.54 million; contributions, $976,000), outraised, outspent, and outcontributed (or matched) the two biggest labor committees, AFL-CIO COPE (receipts, $1.44 million; disbursements, $1.34 million; contributions, $921,000) and UAW-V-CAP (receipts, $1.43 million; disbursements, $1.16 million; contributions, $964,000). Moreover, they outstripped the largest corporate committees, Standard Oil of Indiana, American Family Corporation, and the International Paper Company, by a factor of six to one or greater. Standard Oil of Indiana led all corporations with receipts and disbursements of $266,000, and was second only to International Paper in contributions, $155,000 to $173,000.

Several apsects of business PAC growth are of particular interest. Whereas those labor PACs which were not active financially in 1976 maintained a clear leadership position in 1978, there was a substantive turnover in corporate ranks between the two elections. Perhaps even more important than the switch in corporate identities is the change in the order of magnitude of corporate PAC activities. Whereas in 1976 only nine company PACs had receipts and expenditures exceed $100,000, by 1978 twenty-eight companies were in this category and six had receipts or expenditures exceeding $200,000 (two companies, Standard Oil of Indiana and American Family Corporation exceeded $260,000 in each category). In 1976 only one firm (General Electric) contributed above $100,000 to congressional candidates. By 1978 ten companies exceeded or were within a hairsbreadth of that amount, and the two leaders both contributed more than $150,000 to candidates. Union-related committees have also gotten somewhat larger in campaign 1978. In 1976 the forty-two union committees had receipts in excess of $100,000; by 1978 the figure had increased to forty-seven. Unlike the corporate situation, however, there was only one instance of turnover from 1976 to 1978 among unions with the ten largest PACs (in terms of contributions). It is thus no surprise that between 1976 and 1978, corporate contributions to congressional candidates more than doubled ($4.3 million to

$9.8 million), while labor's contributions rose by only 24 percent ($8.2 million to $10.3 million). Business-related associations also increased in size between 1976 and 1978. As a result of inadequacies in FEC data for that year, it is virtually impossible to calculate precisely the aggregate contributions by all business-related PACs in 1976. The $7.1 million figure (for all races) utilized by Common Cause unquestionably underestimates the amount; $10 million plus is probably closer to the mark. By 1978 aggregate business contributions had risen to an estimated $17.3 million. Whatever measure one uses, it is apparent that business-related PACs (both corporate and noncorporate) played a far more important role in 1978 than they had in any previous election.[30]

Although in 1976 and 1978 labor and business-related PACs demonstrated a clear propensity to be what management scientists term "risk averters," giving money predominately to incumbents rather than to challengers or candidates in open races, we can detect some differences in the contribution patterns in each group between the two elections biennia.[31] In 1978 labor gave $6.1 million (59 percent) to incumbents, $2.2 million (21 percent) to challengers, and $2 million (19 percent) to candidates in open races. Democrats received 94 percent of the monies. Reduced support of incumbents from 1976 (a sizable number of Democratic congressional incumbents did not seek reelection in 1978) was virtually matched by an increased level of support for candidates in open races. In 1978 the pattern of corporate contributions was virtually identical to labor's, favoring incumbents ($5.8 million, 59 percent) over challengers ($2 million, 21 percent) and candidates in open races ($2 million, 20 percent). Republicans received $6.1 million (63 percent) and Democrats $3.6 million (37 percent), an increase of 6 percent for Republicans from 1976. A *National Journal* analysis of preliminary FEC data for 1978 indicates an interesting aspect of corporate contribution patterns for that campaign.[32] Through September 30, 1978, corporations divided their contributions fairly evenly between Republicans ($2.5 million, 53 percent) and Democrats ($2.2 million, 47 percent), but heavily favored incumbents ($3.4 million, 72 percent) over nonincumbents and open-race candidates together ($1.3 million, 28 percent). In the last month of campaign, however, corporate PACs contributed heavily to Republicans ($2.9 million, 71 percent) over Democrats ($1.2 million, 29 percent) and slightly favored nonincumbents and candidates in open races ($2.1 million, 51 percent) over incumbents ($2 million, 49 percent) as viable Republican challengers and close open races become more apparent.

To summarize, the conventional wisdom of corporate-related dollars being automatically Republican dollars was heartily disproved in both 1976 and 1978. Moreover, while both elections indicated an overall corporate PAC preference for incumbents over challengers and open-seat candidates, the 1978 election suggests an increasing propensity on the part of corporate PAC managers to target monies to nonincumbents, who most frequently are Republicans. Given the increasing frequency of open seats during the past few elections (particularly in the House), corporate PACS are likely to devote even more new resources to challengers and candidates in open races than was the case in 1978 and to be more active in primary races. In short, corporate PACs are expanding their role as risk takers, a move which will favor Republican nonincumbents.[33] Finally, in 1978 noncorporate, nonlabor PACs gave an aggregate $6.4 million (43 percent) to Democrats and $8.6 million (57 percent) to Republicans, and favored incumbents ($8 million, 54 percent) to challengers ($3.5 million, 23 percent) and candidates in open races ($3.4 million, 23 percent).[34]

Several caveats are necessary before we leave the subject of business and labor behavior during campaigns 1976 and 1978. First, along with establishing PACs, corporations, unions, and other groups may advocate (in communications to their stockholders, managerial personnel, and members) the support or defeat of particular candidates. While labor has made extensive use of internal communications, corporations have done little in this regard.[35] Second, the above figures do not include labor union spending for registration, get-out-the-vote drives, candidates, logistical support, and general political education; activities that benefit labor-endorsed candidates and are considered by many political observers to be even more important to a candidate's campaign than direct financial contributions.

Stipulations filed by three large unions in conjunction with the Republican National Committee's suit against the FEC give some indication of the nature and extent of such labor activities in 1976. The International Association of Machinists (IAM) report that aggregate outlays in 1976 by all IAM constituent units (international, state, district, and local) totaled $351,000 for internal communications, voter registration, and get-out-the-vote activities.[36] The United Autoworkers' stipulation indicates that for the period July 15, 1976, to November 2, 1976, the union spent $1.6 million on such activities.[37] AFL-CIO COPE indicate that state and national COPE expenditures post-July15, 1976, for membership identification, registration, get-out-the-vote, presidential endorsement and publicity, and overhead totaled nearly $3 million.[38] All told, in 1978 labor probably spent nearly $20 million for these items. While thus far some national business groups (such as BIPAC and the Chamber of Commerce of the United States) and an occasional corporation have undertaken serious political education efforts, business has in general done very little in voter registration, get-out-the-vote, and non-candidate-related internal political communications and in-kind support of political candidates. While corporate and other business-related associations are exploring such involvements for 1980, it is unlikely there will be extensive business activity in these areas before 1982 or 1984 by which time their PAC solicitation and contribution programs will have matured, thereby permitting greater resources and opportunities to experiment with new forms of electoral involvement.

Third, both labor and business have not made much use of independent expenditures to advocate the election or defeat of a candidate. Such expenditures which can be used to buy media time or space and to promote or oppose a clearly identified candidate are independent as long as they are not made with the "cooperation, . . . prior consent, . . . in consultation with or at the request or suggestion" of a candidate or his agent.[39] To date, however, most corporate and business-related PACs have eschewed the independent expenditure route. Many candidates are leery of independent advertising on their behalf (the content and timing of which they cannot control legally) and actively discourage prospective independent expenditures, particularly by business.

Assessing the PAC Phenomenon

Analysis of the welter of data concerning business and labor PACs contained in the preceding pages yields several useful insights. First, the figures reveal impressive growth in both numbers and financial activity by PACs, particularly those of a business-related nature, and suggest that whereas on the labor side the growth opportunities are limited, the market potential for corporate and other business-related PAC formations and expansion is virtually unlimited. Second, the data indicate that among both unions and corporations there is a positive relationship between the size of an organization and both its propensity to form a PAC and the level of its PAC activity in terms of receipts, expenditures, and contributions. The larger the corporation or union, the more likely it is to have a vigorous PAC. For example, 70 percent of *Fortune*'s top 100 industrial companies operated a PAC during the 1978 election, while only 6 percent of the firms that *Fortune* ranked 900 to 1,000 had a PAC during that period. Third, the data appear to support the view that in addition to organizational size in the case of corporations and other business-related groups, another key factor influences PAC formation and activity: the importance of federal government decisions to the well-being of a firm or industry. Fourth, both business and labor PACs have a distinct and almost identical orientation toward incumbents over challengers, though less so in 1978 than in 1976. Fifth, while labor PACs have overwhelmingly (95 percent) favored Democratic candidates, corporate and other business-related PACs have been decidedly more bipartisan (preferring Republicans to Democrats by approximately a 6:4 ratio). Sixth, figures for 1978 also indicate that ideological groups, particularly those with a conservative orientation, have been the most effective practitioners of the PAC art outstripping both business and labor in their ability to raise monies. Finally, the data suggest that PACs in aggregate and business-related and labor committees in particular still constitute, in numerical terms, a relatively minor source of funds for congressional candidates.

In campaign 1978, the $35.1 million contributed by all PACs amounted to less than 18 percent of the nearly $200 million raised by all candidates, while combined business and labor donations probably provided less than 14 percent of the congressional total. While the essential data regarding the growth of PACs are reasonably clear, the message to be derived from these data is less so. One interpretation of the PAC phenomenon is decidedly negative, viewing the growth of PACs as an important element in making the current system of financing congressional elections a national disgrace.[40] To counteract PAC influence, proponents of this position have favored public financing of congressional elections but have been unable to muster the requisite support in the House. As a second-best alternative, reformers have sought to reduce the amount which PACs can contribute to federal candidates.

In July 1979 a bipartisan coalition of some 120 House members, headed by Representatives David R. Obey (Democrat, Wisconsin) and Tom Railsback (Republican, Illinois) introduced the Campaign Contribution Reform Act of 1979.[41] As proposed, the act would reduce the amount that a PAC could contribute to a House election from $5,000 to $2,500 (primary, runoff, and general elections are each treated separately)—a total of $7,500 per election cycle—and would have limited House candidates from receiving more than $50,000 from all PACs per election cycle. The bill also sought to limit the ability of professional fund raisers and campaign consultants (such as Richard A. Viguerie) to extend credit to their candidate-clients in excess of $1,000 for media advertising or direct-mail fund-raising services. While acknowledging that special-interest groups have a legitimate role to play in the electoral process, a statement by Congressman Obey who introduced the bill opined that "their role must be kept in place to protect the integrity of the congressional process and the rights of all Americans."[42] Obey expresses concern about the relationship between PAC contributions, particularly where the interests of a variety of groups coincide, such as when unions and companies in a given industry have a common position on a given issue. Obey-Railsback passed the House 217 to 198 but not until the proposed contribution limits on PACs were raised from $5,000 to $6,000 ($9,000 in the event of a runoff election) and the amount an individual candidate could receive from all PACS was increased from $50,000 to $70,000 ($85,000 in the event of a runoff). The bill was vigorously opposed by business and conservative groups which argued that it favored organized labor since, while it covers equally direct financial contributions from both business and labor, it doesn't affect such activities as voter registration and get-out-the-vote drives which labor has used to much greater advantage than business. The bill stalled in the Senate as a result of both Republican opposition (including a threatened filibuster) and a crowded legislative calendar.

Not surprisingly, organized labor worried by the explosion in corporate and other business-related PAC activity, strongly supported Obey-Railsback efforts to limit individual and aggregate PAC activity. In the 96th Congress, the same labor representatives who were most instrumental in shaping the PAC provisions in the earlier campaign acts urged that PAC contribution limits be reduced from $5,000 to $2,500 (subsequently raised to $6,000 for a primary and general election) and that partial public financing of House general election races be instituted. Arguably, the effect of nearly halving PAC contributions would be more cosmetic than real, since the great bulk of both labor and business contributions comes in amounts of less than $2,500. Yet it is noteworthy that union PACs gave more than $2,500 and over contributions in 1976 and 1978 than did business. Labor, wisely, is willing to forgo a short-run advantage from maintaining the higher limit in exchange for the longer-run benefit of forestalling the large corporate contributions that could come once company PACs have assembled truly substantial funds. In addition and for very similar reasons, during spring 1979, organized labor strongly backed the two public financing bills (H.R. 1 and S. 623) considered and shelved by Congress. Many labor officials would like to have unions, corporations, and other interest groups wholly out of the business of making direct money contributions, through their PACs, to political candidates and party committees. They foresee that business contributions will eventually outstrip direct labor donations and would prefer, therefore, to restrict business and labor involvement to those activities in which labor has the greatest comparative advantage: voter registration, political research and education, and get-out-the-vote drives.

An independent but not unrelated labor response to the corporate PAC phenomenon in late 1979 was a complaint filed with FEC in October 1979 by the International Association of Machinists against ten of the country's largest corporate PACs.[43] The complaint alleged that "coercive methodology of corporate PAC solicitations from unprotected career employees"[44] violated 4416 (3) of the FECA since it resulted in employee contributions which were neither free nor voluntary as required by the act. The IAM's complaint also contended in separate provisions that if corporate solicitation practices were upheld, the FECA's prohibition of direct union contributions to federal candidates is discriminatory and violates the due process guarantees of the Fifth Amendment. Finally, the IAM contended that the FECA's authorization of financing the costs of corporate PAC operations from company assets violates shareholders' First Amendment rights of free speech and association. The IAM complaint is clearly intended to put pressure on the leading corporate PACs by both subjecting them to close public scrutiny and requiring them to justify the nature and extent of their PAC operations. It is not inconceivable, moreover, that the timing of the complaint was intended to assist passage of the Obey-Railsback bill, then pending consideration by the House.

Another catalyst to efforts by both campaign reformers and organized labor to restrict the role of PACs in election primaries was a report to the Committee on House Administration issued in mid-1979 by the Campaign Finance Study Group with Harvard University's Institute of Politics.[45] One of the three problems stemming from the FECA upon which the Study Group focused was the growth of money channeled through PACs. Decrying the stringent restrictions placed by the FECA on individual contributions ($1,000 per candidate per race,

$5,000 to a single committee per year, and $25,000 overall per calendar year) and on the political parties, the Study Group Report notes that the legislation made candidates increasingly reliant on personal wealth and PACs. The tone of the report reflects considerable concern about PACs terming PAC money as *interested money,* that is, linked to a legislative lobbying agenda.[46] Reliance on PAC funds, opines the report, has led to a nationalization of the sources of money available to candidates, bringing in funds from outside a candidate's state or district (particularly Washington). The growing role of PACs has resulted, moreover, in political monies becoming bureaucratically organized, that is, detached from their source and aggregated in a fashion which renders them unaccountable. Notwithstanding these concerns, the Study Group eschews recommending the imposition of additional limitations on PAC activity. Noting the virtual impossibility of legislatively rolling back the clocks to 1974 to make PACs simply disappear, the Study Group considers that the most probable result of reducing the amount that PACs can contribute to candidates would be to "merely divert, but not stem, the flow of money. Proliferation of PACs, perfectly legal cooperation among PACs, and a rapid expansion in independent expenditures by PACs are the clearly predictable consequences. Considering the combined reasoning which governed the *Buckley* and *Bellotti* decisions, there does not appear to be a legislative remedy for this development that will pass constitutional muster."[47]

The Study Group proposes to resolve (at least in part) the problems of burgeoning PAC contributions by indirect rather than through direct action, urging Congress to raise the individual contribution limit to $3,000 (or better still, $5,000), thereby making congressional candidates less dependent on money derived from PACs.

Prospects for Labor PACs

The PAC phenomenon is still in its infancy. Absent major changes in the overall structure of campaign financing (particularly public financing of congressional races) and limitations on their activities, much more severe than the restrictions embodied in Obey-Railsback, PACs will become an increasingly important factor in funding federal elections. PAC operations in 1976 and 1978 reveal only the tip of a possible iceberg, clearly for corporations and other business-related groups but to some extent even for labor. Let us look briefly at the union PAC potential, which has rather different prospects.

The number of union PACs is *not* going to increase much from the 303 which functioned during 1976. Indeed, the number of labor PACs active during the 1977-1978 election declined to 281. Politically active unions have, in the main, been operating PACs for years. Moreover, union PAC activity tends to be highly concentrated. In 1978 the ten largest union PACs contributed slightly over half of all monies which labor gave to congressional candidates. Eight of the ten PACs were affiliated with the top twenty-five unions ranked on the basis of size; one (Seafarers International) was the forty-fifth-ranking union, and only one (Marine Engineers) had less than 100,000 members.[48] Of the 303 labor committees active in 1976, 42 with receipts or expenditures of over $100,000 raised and spent 82 percent of labor's funds. If another twenty-one union PACs with receipts or expenditures of $50,000 to $100,000 are added in, we have nearly 90 percent of the union total. The remaining PACs represent either small unions or affiliates of large international unions (and as such were subject to the single contribution limit).

It is illusionary, therefore, to look for growth in the number of labor committees. Labor's best opportunity for increasing its pool of voluntary political dollars lies in developing more productive fund-raising techniques. Many unions still rely primarily on raffles, potluck dinners and other fund-raising events, contests, and the like to raise money for their PACs. Very few unions, whether AFL-CIO affiliated or independent, have average contributions of a dollar per worker per year. If even that small amount were collected from each unionist, organized labor would raise some $19.4 million annually or nearly $39 million biennially from its U.S. members; this would almost double the amount generated by labor for campaign 1978 ($19.8 million) and would enable unions to contribute some $20 million directly to congressional candidates. Some unions are beginning to use payroll deductions (checkoffs) to increase their per member annual yield where this method is available through either reciprocal rights or collective bargaining. Others are considering computer-based direct-mail campaigns among their members, a technique used to good advantage by conservative groups and by the National Republican Congressional Campaign Committee. It is assumed that labor will maintain unabated (financed largely out of treasury funds) its political research and education, voter registration, and get-out-the-vote activities which constitute its political forte. A number of unions are beginning to make use of their members' particular expertise to assist candidates. For example, the Communications Workers of America whose members perform a multitude of roles within the Bell and other telephone systems are concentrating on aiding candidates and other unions in organizing and staffing important telecommunication tasks within elections. Accordingly, notwithstanding its declining membership and some evidence of organizational malaise, organized labor's constituency and resources remain sufficiently large, and its political expertise runs sufficiently deep to give the union movement a viable political base, at least in the short run.

The far more interesting view of the PAC iceberg is, of course, on the corporate side. The market potential for company-PAC formation is virtually untapped, even if only the very largest firms are considered to be PAC prospects. As we have seen, in the 1978 election only 24.4 percent of the top 1,000 *Fortune* industrials operated PACs. The vast majority of corporate PACs which are already functioning have ample opportunity, moreover, to increase the size and scope of their operations. Given the trend and the potential, there is no gainsaying that company committees give every promise of continuing to increcase in number and in the magnitude of their funds. In campaign 1980, there could be over 1,000 corporate PACs operating with aggregate receipts of $25 million to $30 million and contributions of $15 million to $18 million. Nearly 950 are operating as of December 31, 1979. By 1982 the number of corporate PACs could reach 1,250, with receipts of $40 million to $50 million and contributions of $25 million to $30 million to congressional candidates (assuming the present disbursement/contribution ratio does not change drastically). My research suggests, moreover, that at least some companies will begin to undertake new forms of electoral involvement such as automatic payroll-deduction plans, nonpartisan registration and get-out-the-vote drives, and internal political communications among managerial-level employees and shareholders).

These projections of future corporate PAC operations are conservative estimates and not reckless speculations. Similarly, business-related (but noncorporate) associations are likely to increase both the size and activity of their PAC operations and to explore more vigorously independent expenditures and in-kind contributions. The pool of potential PAC registrants among associations is large. For example, of approximately 4,100 trade and professional associations nationwide, an estimated 1,500 are currently headquartered in Washington, D.C., alone. An increase of noncorporate, nonlabor PACs to nearly 1,000 for campaign 1980 is quite possible.

Conclusion

Social scientists have long noted that political phenomena frequently have unintended and even paradoxical effects. Labor's policy during the 1970s of seeking legitimization of the PAC mechanism, together with a liberalization of the legal constraints surrounding its electoral activities provide a case in point of such unintended consequences. During the 1970s labor clearly fought for and won the legislative mandate to establish PACs and to engage in other forms of partisan and nonpartisan political activity. Labor's legislative successes had, however, a direct and beneficial effect upon business. The rapidity and effectiveness with which the business community has embraced the PAC mechanism was not anticipated by organized labor's political leadership. Nor did the leadership foresee that the growth of business PACs would erode labor's position by providing alternative sources of funding for both congressional incumbents and challengers (particularly Democrats) who in the past were heavily dependent upon union monies for their campaigns. While it is premature to suggest that labor's overall electoral effectiveness vis-à-vis business has been fatally impaired, it unquestionably has been weakened.

A critical aspect of the aggregate effect of FECA's three rounds during the past decade has not, I believe, been appreciated by either those who see business electoral behavior during the 1970s as so much déjà vu, or those who fear that we are rapidly approaching (if we have not already reached) a point of business hegemony within the American electoral process. For example, Michael J. Malbin, a political scientist associated with the American Enterprise Institute for Public Policy Research and an astute observer of the PAC scene, has argued that campaign funds emanating from business have not really increased in either an absolute or a relative sense, that PAC monies constitute primary funds which business had previously channeled into the campaigns either directly or subrosa through company or associational officials or through various forms of in-kind or otherwise masked contributions, some of which were patently illegal.[49] PAC monies are in essence "old wine in new bottles" (my language, not his). Malbin concludes that "as fast as the funds from business PACs have poured in, fund raisers have been able to find other ways to raise money, and thus keep the PACs proportionately in their place.[50] He argues, moreover, that the PACs and the connection between their gifts and congressional policy are a good deal less significant than we have been led to believe. He asserts further that

> *In light of the record, it is hard to justify the notion that campaign gifts, particularly ones from the more broadly based labor corporate or ideological PAC, are a special-interest group's downpayment for future special benefits. Some associations may think that way, but associations have become a decreasingly important part of business giving as corporate PACs have grown* (emphasis in the original).[51]

Malbin does not believe that at present a problem truly exists with regard to business and labor electoral involvement. While recognizing the potential for business-related and particularly ideological or single-issue PACs to become increasingly important factors in electoral politics, he concludes that far greater dangers exist in the erosion of the parties and the appeals to extremist positions by the new breed of direct-mail political fund raisers and the candidates they support in a manner which further polarizes the political process.

Although Malbin's admonitions are important, they are somewhat to the side of the mark. The essential point is that even if the sums contributed by American businesses to election campaigns have remained constant, the *process* by which corporations and other business-related groups raise and expend campaign monies has changed fundamentally. While in the past, fund raising and expending activities were largely ad hoc, informal, and unsystematic, today such efforts have become *institutionalized* within companies and in the hands of staff professionals (usually in public affairs positions) who serve on an ongoing basis as the organizational focal point for electoral activities. These PACs were visible to officeholders, prospective candidates, and party officials, and to each other, and have become ports of call for office seekers and fund raisers as well as an enabling mechanism for more effective coordination among business groups. In summary PACs provide an exceedingly efficacious device by which corporations and business-related associations can organize and institutionalize their electoral activities. Thus far for most companies, these efforts are still at a rather rudimentary stage of development and have been devoted to raising funds, primarily from upper-level management employees, for contributions to individual campaigns. Current contribution activities are being expanded to include lower-level administrative and executive personnel and, in the case of some companies, shareholders. Moreover, as labor has so well demonstrated for years, PACs constitute an idea way for coordinating within a company a wide range of grass-roots political activities which utilize human and other organizational resources found in a corporation and are so valuable in an election campaign. During the 1980s, we are likely to experience intensified corporate efforts to enlist increased political participation by employees, shareholders, and even retirees. The result could be a much more comprehensive and extensive business electoral involvement than we have experienced in the past.

A second point warrants mention. In addition to institutionalizing electoral activity within the corporate entity through the PAC, the campaign reform legislation of the 1970s has legitimated that activity both within the firm and in the greater community. Electoral politics, so to speak, has come out of the corporate closet and is now acknowledged as a legal and appropriate activity for business. Such enhanced status, together with a clearly defined legal mechanism—the PAC—make it possible for companies to enlist political participation among corporate personnel (who might otherwise be reluctant); encourage other firms to increase their electoral involvement by establishing PACs, thereby keeping up with the Joneses; and in general undertake political activity with a heightened sense of rectitude and purpose. Increased political legitimacy coupled with the institutionalization of politics within the organizational framework has the potential to raise the level of corporate political action both quantitatively and qualitatively. In short, even if to date there has been little change in the degree of business electoral involvement when measured in terms of dollars infused into political campaigns, there has been a fundamental change in kind with regard to that involvement. Whether one considers this institutionalization and legitimization of business electoral activity and its consequences as beneficial or deleterious to the electoral process depends upon one's normative stance, and, to a large degree, upon whose ox is being gored. At the very least, however, it is preferable for such political action to be conducted openly and, as a consequence of the rigorous disclosure requirements of the federal election laws, under public scrutiny rather than, as in the past, to take place within the murky netherworld of illicit corporate activity. Our recent unhappy experience in the 1972 election in which the Finance Committee to Reelect the President (Nixon) demanded and obtained laundered (and illegal) campaign contributions from a number of the nation's largest companies underscores this point dramatically.

PACs have played a useful role in the American electoral process during the 1970s by encouraging and facilitating collective political participation by persons associated with a wide variety of economic, professional, ideological, and other social-interest groups. Special interests, including business and labor, have a legitimate role to play in electoral politics.[52] There can, however, be difficulties arising from their participation. It is important to note that although business and labor are usually cast in the role of electoral competitors, frequently they in fact share sufficient political interests so that cooperation rather than competition can characterize their behavior. Joint business-labor geographical concerns or industry needs, or the opposition of a common foe (a militant environmental group) can mean coalition politics in which the two most powerful coalitions themselves join together. The combined (and successful) efforts in 1979 by Chrysler Corporation and the United Auto Workers to obtain federal assistance for the nation's third largest automaker constitute a recent illustration of this phenomenon. Arguably, one of the greatest potential challenges to the integrity of American electoral politics could arise from excessive harmony between powerful business and powerful labor. While thus far this has not been a serious problem, it could conceivably become one in the future and accordingly bears watching. A second factor requiring scrutiny is the effect of the PAC phenomenon on union members and on corporate employees and shareholders. It will be recalled that throughout this century, in addition to preserving the integrity of the electoral process and those chosen by it from subversion by economic interests, public regulation of corporate and labor involvement has been motivated by a desire to protect individual union members and shareholders from political pressure by their organizations.

Under the FECA reforms, the union member or corporate employee is probably no worse and possibly somewhat better off now than before. The antireprisal provisions in the present federal legislation are likely to be prophylactic for the average union member or middle-level corporate manager who does not wish to participate in the political activities of his organization. The higher one goes in a firm or union, however, the less useful are these statutory safeguards. Upper-level business and labor officials still undoubtedly face subtle peer pressures and psychological arm-twisting. An additional pressure on potential $200+ PAC contributors arises from FECA disclosure requirements. Despite good faith assurances by their organizations that all contributions activity will be kept from the eyes of organizational superiors, corporate managers and union officials are aware that a record of their contributions (or noncontributions) is available for all to see in the FEC's open records; thus the FEC safeguards are hardly fail-safe in protecting their intended beneficiaries. While to date there have been no formal allegations of patently illegal activity by a labor union, the complaint filed with the FEC by the IAM in late 1979 will provide an opportunity to examine closely the solicitation activities of the largest corporate PACs to see whether they violate the spirit if not the letter of the law.

And what of the future? The answer to whether the present regulatory framework governing the involvement of unions, corporations, and other business groups requires a drastic overhauling lies directly in the hands of business organizations and labor unions which operate PACs. Undoubtedly, the most important rationale underlying public regulation of corporate and labor electoral involvement has been to ensure that the power of wealth does not run roughshod over the people's will. As we have seen, in 1978 business and labor PACs provided $35.1 million (approximately 18 percent) of the $199 million raised by congressional candidates. This percentage does not, in my view, amount to excessive interference in the political process or present the kind of threat to the body politic that would justify dramatic new regulation of business and labor electoral behavior. I do not include in the category of "dramatic new regulation" the proposal in Obey-Railsback to reduce the amount which a PAC can contribute to an individual candidate from $10,000 per election cycle to $6,000. I have considerable reservations, however, concerning the imposition of limits on the aggregate amount that a candidate can accept from PACs, particularly the ceiling of $70,000 established in the pending Campaign Contribution Reform Act. Recent studies by Gary C. Jacobson demonstrate convincingly that limiting the amounts which candidates can spend in an election favors incumbents over challengers.[53]

We shall not reach the point of necessitating a substantial revamping of the current legislation unless business or labor badly overplays its hand by misuse or overuse of its PACs. If PACs become too successful—that is, if in aggregate they become a disporportionate source of funds for congressional candidates (for my taste, in excess of 25 percent) which could be the case if business-related PACs should raise $50 million per election biennium and contribute half that directly to federal candidates, and labor PACs achieve their $40 million goal—we will indeed reach a point where too much campaign money originates from these sources. This would be particularly the case if business and labor contributions were largely reinforcing: supporting incumbents (though different ones to some extent), doing little for challengers, and thereby perpetuating the congressional status quo. Similarly, if a substantial number of our largest corporations begin to generate PACs with receipts in excess of $250,000 to $300,000 and contributions of $140,000 to $200,000, so that big business becomes too identified in the public mind as a source of political campaign funds or otherwise become too ubiquitously involved with the electoral arena, the limits of public tolerance will likely be exceeded. This will be the case particularly if, as I suggested, organized labor fails to keep pace financially and thereby exercise at least a semblance of countervailing power in electoral politics. Finally, if a sizable number of business or labor PACs coordinate closely their efforts to throw large sums of money into the final stages of an election in an effort to tip close races or to "get" candidates appearing on an "enemies' list," allegations of excessive PAC power would be justified.

For now, at least, these future possibilities are in the realm of speculation. What is certain, however, is that American unions enter the 1980s with a position in federal electoral politics considerably weaker than they occupied a decade earlier. Three rounds of federal electoral legislation during the 1970s have provided a congenial regulatory setting for expanded electoral involvement by corporations and other business-related groups. Indeed, the very statutory changes sought by organized labor in 1971, 1974, and 1976 to satisfy its short-run political needs have had the ironic effect of sowing the seeds which have borne the specific fruit which throughout the twentieth century the union movement has sought to prevent business from harvesting: enhanced electoral effectiveness. Compounding the irony is the fact that American business' expanded role in federal elections during the 1970s has come about largely through its adoption of labor's long-standing and favorite mechanism in electoral politics, the PAC. In short, the PAC phenomenon of the 1970s has been primarily a business phenomenon, a phenomenon moreover which has eroded the electoral position of American labor.

Notes

1. The AFL-CIO and the United Autoworkers have become particularly staunch proponents of both public financing of congressional elections and legislation to limit the amounts which PACs can contribute to congressional candidates. In both the 95th and 96th Congress, labor representatives advocated partial public financing of House general election races (H.R. 5157, 95th Cong., 1st sess.; and H.R. 4970, 96th Cong., 1st sess.) and the reduction of PAC contribution limits (H.R. 11315, 95th Cong., 1st sess.; and H.R. 623, 96th Cong., 1st sess.). See commentary by Ben Albert, AFL-CIO Committee on Political Education (COPE) information director, in "Letters," *Regulation* (September/October 1979): 2.

2. In late 1979, the International Association of Machinists filed a complaint with the Federal Election Commission against eleven corporations (including companies with the ten largest PACs) alleging that these firms were violating the Federal Election Campaign Act and the United States Constitution by using coercive means to secure involuntary employee contributions to company PACs and were utilizing corporate assets contributed by shareholders without the assent of the latter, thereby violating their constitutional rights. Complaint against Operations of Corporate Political Action Committees including: Dart Industries, Eaton Corporation, General Electric, General Motors, Grumman Corporation, International Paper, Standard Oil of Indiana, Union Camp, Union Oil, United Technologies, and Winn Dixie, by the International Association of Machinists et al. before the Federal Election Commission (Cited hereafter as IAM v. Dart Industries et al.)

3. Common Cause, for example, vigorously opposed in 1974 the successful effort by the AFL-CIO to amend 18 U.S.C. sect. 611 (1974) of the U.S. Criminal Code to permit corporations and labor unions which are government contractors to establish PACs. See Comments by Fred Wertheimer, senior vice-president of Common Cause, in "Letters," *Regulation* (September/October 1979): 3.

4. This chapter does not cover the activity of ideological or single-issue PACs which have flowered during the 1970s.

5. 57 Stat. sect. 167 (1943).

6. 61 Stat. sect. 159 (1947).

7. 18 U.S.C. sect. 610 (1951).

8. See United States v. C10, 335 U.S. 106 (1948), pp. 115, 134-135, and United States v. International Union UAW, 352 U.S. 567 (1957), p. 570.

9. The information about COPE is drawn from Alexander Heard, *The Costs of Democracy* (Chapel Hill: University of North Carolina Press, 1960), pp. 178-208; Fred Greenstone, *Labor in American Politics* (New York: Alfred A. Knopf, 1969), pp. 39-80; J. Cottin and C. Culhane "Committee on Political Education," in G. Smith, ed., *Political Brokers: Money, Organization, Power and People* (New York: Liveright/National Journal, 1972); and Harry M. Scoble, "Organized Labor in Electoral Politics," *Western Political Quarterly* 16, no. 3 (September 1, 1963): 666-685. Also see Charles Rehmus, Doris B. McLaughlin, and Frederick H. Nesbitt, eds., *Labor and American Politics: A Book of Readings* (Ann Arbor, Mich.: University of Michigan Press, 1978).

10. Pre-1971 business electoral involvement is discussed in Heard, *Costs of Democracy*, pp. 98-135; Herbert E. Alexander, *Money in Politics* (Washington, D.C.: Public Affairs Press, 1972), pp. 137-182; and Edwin M. Epstein, *Corporations, Contributions, and Political Campaigns: Federal Regulation in Perspective* (Berkeley, Calif.: Institute of Governmental Studies, 1968).

11. For a detailed discussion of the pertinent legislative and judicial events of the 1970s which have established the present legal framework for labor and business in federal electoral politics, see Edwin M. Epstein, "Corporations and Labor Unions in Electoral Politics," *Annals of the American Academy of Political and Social Science* 425 (May 1976): 33-58; idem, "Labor and Federal Elections: The New Legal Framework," *Industrial Relations* 15, no. 3 (October 1976): 257-274; and idem, "The Emergence of Political Action Committees," in Herbert E. Alexander, ed., *Political Finance: Sage Electoral Studies Yearbook*, vol. 5 (Beverly Hills, Calif.: Sage Publications, 1979), pp. 159-197.

12. L. 92-225, 86 Stat. 3 (1972) sect. 205, codified as 18 U.S.C. sec. 610 (1972).

13. Pipefitters Local Union No. 562 v. United States, 434 F. 2d 1127 (CCA 8th, 1970).

14. In United States v. C10, 335 U.S. 106 (1948) and United States v. UAW, 352 U.S. 567 (1957), the Supreme Court had given broad latitude under

the First Amendment to a union's right to communicate with its members.

15. 407 U.S. 385 (1972). For additional discussion of the Pipefitters decision see Epstein, "Corporations and Labor Unions," pp. 40–42.

16. Common Cause v. TRW, Inc., C.A. 980-72 (D.D.C. 1972). Common Cause dropped the suit when TRW agreed to dissolve its PAC. Accordingly, the incompatibility of 610 and 611 was never judicially determined.

17. L. 93-443, 88 Stat. 1263 (1974). For background and details of the 1974 act, see David W. Adamany and George E. Agree, *Political Money: A Strategy for Campaign Financing in America* (Baltimore, Md.: Johns Hopkins University Press, 1975).

18. 18 U.S.C. sect. 611 (1974).

19. Federal Election Commission, Advisory-Opinion 1975-23, *Federal Register* 40, no. 233 (December 3, 1975): 56584-56588.

20. 424 US 1 (1976).

21. L. 94-283, 90 Stat. 475 (1976).

22. For more extensive examinations of the data pertaining to PAC activities, see Edwin M. Epstein, "Business and Labor under the Federal Election Campaign Act of 1971," and Michael J. Malbin, "The Extent and Significance of PAC Growth," in Michael J. Malbin, ed., *Parties, Interest Groups and Campaign Finance Laws,* (Washington, D.C.: American Enterprise Institute for Public Policy Research, 1979); Edwin M. Epstein, "The Business PAC Phenomenon: An Irony of Electoral Reform," *Regulation* 3, no. 3 (May/June 1979): 35–41; idem, "Political Action Committees."

23. FEC Press Office, "'PAC' Growth–1974 to Present" (n.d. 1980).

24. Figures are drawn from the following FEC sources: press release, May 10, 1979, p. 3; press release, June 29, 1979, p. 3; FEC Disclosure Series, 1976 Campaign, nos. 6, 7, 8, and 9 (1977), and no. 10 (1978).

25. Common Cause, *1976 Federal Campaign Finances, Vol. 1: Interest Group and Political Party Contributions to Congressional Candidates* (Washington D.C.: Common Cause, 1977), pp. VI–VIII.

26. See sources cited in note 24.

27. U.S. Department of Commerce, Bureau of the Census, *Statistical Abstract of the United States* (Washington: U.S. Government Printing Office, 1977). p. 561.

28. FEC press release, May 10, 1978, p. 7, and *Fortune* Directories, May 7, June 18, and July 16, 1979.

29. See FEC, press release, May 10, 1979, p. 3.

30. See sources cited in note 24.

31. The data in this section are drawn from FEC Disclosure Series, 1978 Campaign, nos. 8 and 10, and FEC press release, May 10, 1979, p. 3.

32. Glen Maxwell, "At the Wire, Corporate PACs Come through for the GOP," *National Journal* (February 3, 1979): 174–177.

33. Business organizations and conservative groups have been vigorously urging corporate PACs to increase their support for business-oriented candidates, typically Republican challengers and candidates in open races, over incumbents, who are usually Democrats. See memorandum to Washington Representatives by Clark McGregor for the National Chamber Alliance for Politics (October 17, 1978), accompanying letter by Donald M. Kendall, chairman of Pepsico (October 17, 1978), and *The New Right Report* (Viguerie Communications Corporation), vol. 8, no. 11 (July 31, 1979), "Corporate PACs: Some Improvements, but Still Disgraceful," pp. 1-4 and attachment.

34. FEC press release, May 10, 1979, p. 3.

35. Only 27 of 164 companies responding to a Public Affairs Council survey indicated they had communicated on partisan matters in 1978. Public Affairs Council, *PAC News*, issue no. 7, July 27, 1979 (Washington, D.C., Public Affairs Council), p. 2.

36. Republican National Committee, et al v. Federal Election Commission, 78 Civ. 2783 (LPG) (S.D.N.Y.) IAM Stipulation, p. 13.

37. Ibid., (USDC) (S.D.N.Y.) UAW Stipulation, p. 5.

38. Ibid., (S.D.N.Y.) AFL-CIO Stipulation, Appendix A.

39. FEC Regulations 109.1 (a), 11 FRC 109.1 (a) (1977).

40. Prepared statement of Congressman John B. Anderson (R. Ill.), U.S. Congress, House of Representatives, Committee on House Administration, *Hearings on H.R. 1 and Related Legislation,* 96th Cong., 1st sess., March 1979, p. 216.

41. H.R. 4970, 96th Cong., 1st sess., 1979.

42. Statement of Representative David R. Obey on the "Campaign Contribution in Reform Act of 1979," July 26, 1979. Press release issued by the Democratic Study Group, p. 2.

43. IAM v. Dart Industries et al.

44. Ibid., p. 15.

45. The Institute of Politics, John F. Kennedy School of Government, Harvard University, *An Analysis of the Impact of the Federal Election Campaign Act, 1972-78: A Report by the Campaign Finance Study Group to the Committee on House Administration of the U.S. House of Representatives* (May 1979). Most of the Study Group's useful critique of PACs is the product of Xandra Kayden's research and analysis.

46. Study Group, *Report,* pp. 1-8.

47. Ibid., pp. 1-9.

48. U.S. Department of Labor, Bureau of Labor Statistics, *Directory of National Unions and Employees Associations,* 1975 (Washington: U.S. Government Printing Office, 1978), pp. 17-46.

49. Michael J. Malbin, "The Business PAC Phenomenon: Neither a Mountain nor a Molehill," *Regulation* 3, no. 3 (May/June 1979): 41-43; idem, "Campaign Financing and the Special Interests," *The Public Interest,* no. 56 (Summer

1979): 21–42; idem, "The Extent and Significance of PAC Growth."

50. Malbin, "Business PAC Phenomenon," p. 43.

51. Malbin, "Campaign Financing," p. 36. Reprinted with permission of the author from *The Public Interest,* no. 56 (Summer 1979): 36. Copyright 1979 by National Affairs, Inc.

52. I have developed this position with regard to corporations in Edwin M. Epstein, *The Corporation in American Politics* (Englewood Cliffs, N.J.: Prentice-Hall, Inc., 1969), pp. 304–314.

53. Gary C. Jacobson, "The Effects of Campaign Spending in Congressional Elections," *The American Political Science Review* 72, no. 3 (June 1978): 469–491; idem, "Public Funds for Congressional Campaigns: Who Would Benefit?" in Herbert E. Alexander, ed., *Political Finance: Sage Electoral Studies Yearbook,* vol. 5 (Beverly Hills, Calif.: Sage Publications, 1979), pp. 99–127.

12 The Political Economy of Worker Pension Funds

Laura Katz Olson

Introduction

Increasing concentration of economic power within the corporate sector as well as the extent of corporate control by top wealth-holders, or a few financial institutions such as commercial banks and insurance and investment companies, have been subjects of intense scrutiny and debate in recent years. Of particular concern is the impact of financial power on public policy.

In January 1978 the late Senator Lee Metcalf's Subcommittee on Reports, Accounting and Management revealed that "power to vote stock in 122 of the largest corporations in America, whose common stock represents 41 percent of total market value of all outstanding, is concentrated in 21 institutional investors."[1] The most influential is Morgan Guarantee Trust Company of New York (major identified stockholder in twenty-seven and among the top five in fifty-six corporations), followed by Citibank (major identified stockholder in seven and among the top five in twenty-five corporations).

Similarly, in a study of the 200 largest nonfinancial corporations, which include industrial, transportation, utility, and retailing firms, David Kotz found that financial institutions exercise predominant control over key U.S. enterprises. In defining control as participating in the selection of management, successful pressuring for key policies, or serving directly in a decision-making capacity, Kotz discovered that financial institutions dominate nearly 40 percent of the major corporations. He concluded that "leading bankers have the power to determine or influence the allocation of capital over a significant portion of the economy, and to influence many other aspects of corporate behavior as well."[2] Moreover, bank trust departments' control over these top companies appears to be growing.[3] Big institutional investors, which require as little as 1 or 2 percent of a company's stock to influence the board of directors and its politics, presently hold a substantial percentage of total corporate securities.

It is the assumption here that the role of private and public worker retirement trusts in providing power for these financial institutions has not been addressed adequately. Pension fund assets, which were relatively inconsequential prior to 1960, and their investment in and share of total corporate equities minor, have grown to approximately $470 billion by the end of 1978. Retirement systems currently own between 20 and 25 percent of all common stock and 40 percent of corporate bonds, and, for the most part, these securities are managed and controlled by large institutional investors, particularly commercial banks. Often pension funds represent the second and sometimes primary source of discretionary stock-voting power for these money managers. Despite these factors, there has been limited analysis of or even information on the political, social, and broad economic implications or impacts of worker pension funds.[4]

My central argument is that pension funds have the potential to serve political and social purposes, in addition to economic goals, as well as to promote fundamental change in society. Since this is an exploratory analysis it shall be my aim to gather, present, and clarify basic information on public and private retirement trusts and to propose alternative issues for future research in this critical area.

In their analysis of and interest in public and private pension assets, scholars, employers, and even employees have tended to emphasize narrow economic issues rather than point to the potential economic and political power of these trusts. The focus has been on investment performance, with portfolios assessed solely with regard to rates of return. The objective, especially for private trust funds, has been to earn the best possible rate of return, within acceptable risk parameters, that investment opportunities permit. The Employee Retirement Income Security Act (ERISA) of 1974, the only major legislation focusing on private pension funding, has, among other things, reinforced these tendencies.[5]

These limited concerns have diverted attention from some critical questions. Even during consideration of ERISA several significant social, political, and economic issues were sorely neglected.[6] Despite this apparent indifference, workers' funds have become a major source of capital in the American economy, and as such have been used to help create and sustain practices that adversely impinge on workers themselves. Private and public retirement trusts have contributed to (1) the increasing power of financial institutions; (2) growth of corporate profits that only minimally benefit some plan participants; (3) capital shortages for socially useful investments; and (4) support of corporate enterprises that refuse basic worker demands, such as unionization, and that thwart important community goals.

Characteristics and Financial Aspects of Pension Plans

The two basic types of pension plans explored here are public and private trusts. However, the former will include only state and local retirement systems since the Federal programs tend to be pay-as-you-go systems and invest primarily in U.S. Treasury Securities. Although there are approximately 6,630 state and local pension systems, covering nearly 12.7 million workers, the bulk of the assets are concentrated in a few state-administered funds.

Private plans are composed of both insured and noninsured schemes. Under the insured category, which I will discuss only intermittently, the employer pays premiums to an insurance company which is then responsible for all retirement benefits. Single-employer noninsured plans are the responsibility of a board of trustees appointed by the employer, or occasionally in bargained plans, by the employer and union jointly. The latter arrangement is predominant in multiunion and union noninsured plans, which cover about 20 percent of all private-sector workers. These include mining, construction, wholesale and retail trades, transportation, and other industries in which employment mobility is prevalent and where small firms predominate. Multiemployer and union pension programs are highly concentrated, with the ten largest plans holding over a third of the assets. Similarly, the top twenty-five corporate sponsors hold nearly 25 percent of total single-employer funds.

In recent years, state and local pension trusts have moved away from pay-as-you-go systems and in the direction of funding. Steadily increasing their asset holdings from less than $2 billion in 1940 to $20 billion in 1960, these plans nearly tripled their assets to $58 billion ten years later. Growing by approximately 12 to 15 percent annually, state and local pension systems have amassed $149 billion by the end of 1978.

At the same time, investments have shifted from U.S. government obligations and state and local bonds to corporate securities. As late as 1961 corporate stocks accounted for only 3 percent and corporate bonds for 39 percent while public-sector securities represented nearly 50 percent of investments. By 1976 the former accounted for approximately 74 percent while public-sector obligations represented only 11 percent of total state and local trust portfolios. Moreover, public employee retirement systems have been increasing holdings of equities at a faster rate than other institutional investors. From 1970 to 1978, equity investments alone grew from 13 to 22 percent.

Private pension assets have also had a spectacular pattern of growth. While the noninsured funds rose from less than $2 billion in 1940 to only $6 billion ten years later, from 1955 on they began to vastly outpace state and local systems. Holding $33 billion in 1960, private noninsured trusts amassed $97 billion by 1970 and rapidly approached $202 billion at the end of 1978. In combination with the insured systems, private trusts held a whopping $321 billion.

Prior to the 1950s there were divergent investment priorities. In 1939 corporate stocks represented 20 percent of total assets while 15 percent was invested in federal securities. However, by 1945 the latter represented nearly half of the reserves. Moving from U.S. government securities after the war, the retirement trusts began to acquire corporate equities rapidly and "provided a major remedy to the shortage of equity capital."[7] By 1966 "pension funds had become the largest buyers of common stock, accounting for 10 percent of all transactions."[8] Investment priorities have continued to emphasize corporate securities, though during 1978 private pensions acquired equities at the lowest level in thirty years. At the end of that year, 52 percent of the assets were invested in stock, 39 percent in debt securities (primarily corporate bonds),

6 percent in liquid assets, and the remainder in mortgages, real estate, and foreign government securities. Thus both private and public worker retirement systems contribute substantially to corporate capital markets.

Control of Pension Investments

Existing administrative arrangements for pension funds have tended to buttress the power of financial institutions over the economy. For the most part, covered workers are entitled to a pension if they satisfy specific and often stringent conditions—employees seldom retain decision-making power, through their unions, over investment of the funds. With single-employer noninsured private plans, for example, unions are accorded no roles whatsoever in the administration of the funds. Managers, who tend to be commercial bankers, are appointed exclusively by corporate trustees. Although multiemployer funds have joint union-employer trusteeships, and some union plans are controlled by union officials alone, nearly all these assets are also turned over to institutional money managers, usually with full investment authority and stock-voting rights.

ERISA has tended to strengthen as well as to expand the control of financial institutions over private fund investments. For example, the act mandates fiduciary responsibility for fund investment practices and states that asset managers are responsible for losses sustained from violating the "prudent man" rule. It has been suggested by Burt Seidman, social security director of the AFL-CIO, that this has encouraged turning over even more funds to institutional managers for self-protection.[9]

State and local systems, on the other hand, often have both employee and consumer representatives who influence investment practices. Most of the larger state and local plans place full control over trust assets in retirement or investment boards where employee members constitute a majority in 25 percent and a near majority in another 34 percent of these systems.[10] However, banks and insurance companies serve as custodians or outside managers for a substantial number of public pension systems. While the power of these outside custodians/managers/advisors varies widely, they often obtain full or partial control over the assets. Some observers have even suggested that there is a growing trend among public systems toward hiring outside equity managers and providing them with a high degree of investment discretion and stock-voting rights similar to powers obtained through the private pension systems.[11] Further, "in some cases, one-half of the members of the investment committees (of state and local boards) represent financial institutions."[12]

A 1979 study by *Pensions and Investments* of the top 450 pension managers showed that tax-exempt assets, consisting mostly of employee benefit funds, represented 51 percent of total assets under management by these financial institutions. In fact, from 1977 to 1978, total worker benefit funds managed by institutional investors increased by 9 percent.[13]

Significantly, financial institutions had complete discretionary power over investments for 81 percent of pension trust assets. Full discretionary power increased by 10 percent during 1977 and 5 percent in 1978, indicating that pension trustees are delegating more power to the financial institutions.[14] At the same time they tend to acquire full voting rights over the stocks they purchase for the pension trusts. For instance, Morgan Guaranty exercises proxy voting rights for 98 percent of the retirement funds managed by the bank. What this means, of course, is that the rapid growth of pension assets managed by institutional investors will continue to increase their control over major corporations, and the economy, even further.

Worker pension funds are also concentrated in the hands of a few elite institutions. The top ten money managers alone held $114 billion of the tax-exempt assets. The great bulk of noninsured pension funds are controlled by only about twenty-five banks which held $116 billion. This represents 67 percent of the $172 billion held by the top 155 banks profiled in the *Pensions and Investments* study.[15]

Despite these trends, workers do have the power for regaining some control over their funds and thus power over American corporations through shareholder voting and other methods. Employees in the private sector, for example, could have input into investment decisions of jointly trusteed (multiemployer) and union-sponsored plans. Together with nonprofit organizations, these worker trusts represent nearly 16 percent of total pension assets. If one adds state and local retirement trusts, which include another 32 percent of the total, employees could have an impact on investment decisions for nearly half of all trust fund assets.

Moreover, a strong case can be made for union ownership of and control over all retirement trusts. Although employers tend to finance private plans, it is a fallacy to assume that they bear the major costs. It is generally agreed that contributions to pension plans are in lieu of wages, and this factor is taken into account when collective wage agreements are negotiated. Since the incidence of employer contributions tends to be borne by the employee in the form of decreased salaries, there would appear to be substantial justification for arguing that the funds belong to the workers entirely, rather than to the corporations, and thus should be under worker control. Moreover, 85 percent of state and local workers are required to contribute to their plans and thus a significant percentage of these assets accrue directly through the workers themselves.

Who Gains?

It has been implied that pension systems encourage interlocking interests between employees and employers of different corporations. Since the various trust funds hold equities and bonds in each others' companies, an illusion is created that what is good for each corporation is good for all workers through their retirement trusts. However, the following discussion suggests that current investment practices have had negative consequences for workers and society as a whole. Further, most workers do not gain sufficiently from the retirement benefits themselves.

Current investment decisions made by major institutional investors contribute to the sustenance of corporations that thwart major worker and community goals, including firms that pollute the environment, flagrantly violate civil rights or labor relations laws, provide unhealthy industrial working environments, and engage in other socially injurious practices. Senator Metcalf noted that the present practice of pension administration and control results "in the use of pension money to assist 'notoriously antiunion companies.'"[16] For example, the Ford-UAW retirement plan, one of three United Auto funds, holds a stock portfolio of approximately $873 million. These assets are controlled by large banks which have invested them primarily in twenty-four major corporations. Eleven of these firms are among sixteen categorized by the AFl-CIO as leading antiunion or nonunion corporations.[17]

A 1979 study by Corporate Data Exchange reviewed the stock portfolios of the twenty largest public and 122 major private pension trusts. Out of $44.6 billion in common stock held by the private plans studied, 30 percent was invested in fifty companies that are nonunionized, 5 percent in fourteen firms with serious health and other hazards, 9 percent in twenty-six firms violating equal employment opportunity standards, and 19 percent in thirty companies with significant activities in South Africa.[18] The percentages for the $16.6 billion held by public plans are 26 percent, 7 percent, 8 percent, and 20 percent, respectively. Forty-six percent of total equities held by private and 44 percent of those held by the state and local workers' pension systems support major firms whose activities are counter to the interests of plan beneficiaries and the public interest.[19]

Moreover, since money managers tend to invest nearly all pension assets in the largest corporate firms traded on the New York and American Stock Exchanges they profoundly and adversely affect financial markets. Consequently present investment practices have helped to create a shortage of capital for small and emerging firms.[20] Since the latter tend to be labor intensive, and have been responsible for major gains in employment opportunities, their inability to attract capital exacerbates unemployment in the United States.

Further, as persuasively argued by Jeremy Rifkin and Randy Barber in *The North Will Rise Again,* workers' pension funds have supported the movement of firms and capital from the North and Midwest to the sunbelt region thus contributing to the fiscal crisis, dearth of employment opportunities, and pressures on services in the industrial northern states and localities.[21] The recent uncertainty about the American economy, combined with employer concerns for increasing rates of return, have encouraged money managers to seek new and expanded worker pension investments abroad. By January 1979, 17 of the top

100 pension funds were investing at least some of their assets overseas and 6 others were considering it. International investing of pension assets will increase considerably in the next decade with substantial negative impacts for U.S. labor and American communities.

The rapidly growing asset reserves of state and local systems and the shift from state and local bonds and mortgages to common stocks have helped to create a dearth of available funds for municipal projects thus exacerbating financial problems of cities. Significantly, in procuring equities, state and local governments are relinquishing political control over the utilization of tax dollars. On the average, even self-administered public retirement trusts do not exercise voting rights over corporate policies that their share of stock would entitle them to.

In the past, there were limited efforts to establish union control over pension funds or to encourage socially desirable investments. When John Lewis, president of the United Mine Workers, initiated demands for a welfare and retirement fund during 1945-1946, he also demanded but failed to secure total union control over the assets. Similarly, in 1949 Walter Reuther, president of the United Auto Workers, began negotiations over pension plans. Although he requested union-employer control, the settlement provided for company selection of trustees.[22] In the ensuing year Reuther attempted but failed to influence the managers to invest in housing, day-care, and other socially beneficial projects.[23] However, in general unions have been mostly concerned with immediate worker needs rather than with obtaining control over the retirement trusts. For employees the objective of pension funds has been to maximize retirement benefits and to liberalize qualifying conditions. Unions have also been concerned with ensuring that the company is financially capable of fulfilling pension obligations when they come due, though ERISA has eased this problem somewhat.

However, due to restrictive eligibility requirements and inadequate coverage and level of benefits, even those participating in public and private plans may not gain sufficiently from them. First, there is a considerable difference between covering a worker and actually providing retirement benefits. In 1978 only about 40 million workers, or approximately 50 percent of private-sector employees, were covered by private pension plans. However, in that same year over 75 percent of retired workers were not receiving benefits. One of the reasons is restrictive vesting regulations which have been liberalized only somewhat by ERISA. The legislation does not provide for full and immediate vesting but rather offers three options to private employers. Most plans that had to amend their vesting provisions chose full vesting after ten years of participation under a plan. Yet it has been noted by the Department of Labor that men stay on their jobs an average of only 4.6 years, and women 2.8 years.[24]

Although nearly all state and local workers are covered by pension plans, these public-sector employees are not included under the provisions of ERISA.

Consequently there have been few changes in vesting for these schemes. About 70 percent of state and local plans, covering one-fifth of the employees, fail to meet even the minimum vesting requirements of ERISA.[25] There is also an absence of reciprocity among states; workers who tend to move frequently around the country often lose benefits that have been paid into the funds by state and local employers.

Second, even those employees who are entitled to a private pension may find the replacement of previous earnings quite low, even with social security. In 1975 the average annual private retirement payment was estimated to be about $2,100.[26] Further most plans do not provide adequate protection for the worker's family. By the mid-1970s, a staff report of the U.S. Commission on Civil Rights reported that only about 2 percent of the elderly widows of covered employees were receiving benefits under the private retirement system.[27] ERISA now requires private employers to make available automatically, unless the employee chooses to cancel it, an actuarial equivalent joint-and-survivor option. Under nearly all plans, however, there is no widow's benefit if the husband dies before age 55, even if he is vested. Early widow's benefits after age 55 are not automatic—the plan must have a provision allowing for an early suvivor option, and the husband must be vested, working under the plan at the time of his death, and have a signed agreement stating that he would have taken a reduction in his early retirement pension. Since these various survivor options reduce pensions considerably, many employees, particularly those with low annuities, select the higher benefits, thus leaving the spouse unprotected. Moreover, if the widow is eligible for a benefit, it will be only a small proportion of the pension accrued by the husband if he dies prior to retirement. Critically, even female workers, due to their intermittent employment patterns and, more important, their concentration in industries without pension coverage, tend to be ineligible for pensions in their own right.

Third, for the most part private pensions do not have cost-of-living adjustments, and therefore inflation erodes purchasing power of retirees substantially. Although some employers have added supplementary benefits, these have been sporadic and insufficient. In contrast, nearly 90 percent of state and local workers are in plans that have some kind of cost-of-living adjustments. However, most of these increases are on an ad hoc basis. Nearly half of the retirees have their pensions increased annually, usually at rates below the consumer price index (CPI), while others are partially tied to the CPI. Less than 5 percent of state and local workers participate in plans that are automatically and fully adjusted for inflation.[28]

For employers pension goals focus on augmenting total investment earnings; as the latter increase in value, through market gains, employers can decrease financial commitments to the funds. Jack L. Treynor, a financial analyst, has estimated that reductions in employer contributions can fluctuate from 16 to 30 percent for each 1 percent increase in investment returns.[29] The rewards from superior portfolio performance tend to benefit the plan-sponsoring company. These enterprises have also gained enormously from tax policies relating to the pension systems. Since employer contributions into the pension funds, the accumulated assets, and income earned through investments are tax-exempt, the loss to the national treasury is considerable. Currently the tax subsidy to private plans is approximately $10 billion annually.[30]

Symbolically, while in the United States high investment yields tend to be related to decreased employer input into the funds, in some countries the presumption is that workers will reap the benefits of fund growth. In the Netherlands, for example, as investment earnings rise there has been a tendency to add regular increases for retired workers. While payments in Sweden are not automatically indexed to inflation, occupational pension benefits have kept in line with increases in the retail price index. Further, instead of reducing employer contributions when interest yields are higher than expected, as in the United States, the propensity in Sweden is to pay out bonuses to retired workers.[31] These factors are even more significant if one considers that government controls over investments are quite stringent. For example, only 10 percent of the occupational funds are allowed to be invested in equities or real estate. The rest must be invested in national and local government bonds and securities as well as in other public-sector enterprises. According to Wilson, "This control obviously keeps down the rate of return on the investment of funds, but in spite of this . . . (trustees manage) . . . to keep occupational pension payments in line with rising prices out of interest on investments and surpluses on the business of the fund."[32] The national supplementary pension, also financed by employers, emphasizes public-sector rather than private-sector needs. Half of these funds are in housing credits with the remainder divided into government bonds and loans to municipalities and industrial concerns.[33]

Alternative Investment Goals

The evidence indicates that a few unions and multiemployer funds as well as some nonprofit and state and local retirement trusts have been utilized, though only to a limited extent, for the furtherance of divergent goals, including housing and other socially desirable projects as well as for political and economic power.[34] Moreover, there are some indications that organized labor and a growing number of public leaders are reevaluating their neutral positions on investment practices.

In November 1977 William W. Winpsinger, president of the International Association of Machinists, proposed that his union's retirement assets of over $80 million be withdrawn from its present managers, Manufacturers Hanover Trust Co., since James D. Finley, board chairman of J.P. Stevens, is a member of the board of directors. As is well known, J.P. Stevens and organized labor are at odds over a number of critical worker issues including unionization itself.

Other unions with trusts managed by Manufacturers Hanover, encouraged by Ray Rogers, director of the Amalgamated Clothing and Textile Workers' Corporate Campaign, protested Finley's presence on the board, and also threatened to withdraw their funds' assets. The bank manages over $1 billion in pension funds. By March 1978 Finley was forced to resign from Manufacturers Hanover. A second J.P. Stevens director, David W. Mitchell, president of Avon Products, Inc., was also pressured into leaving his position as a director of the bank as well as that of J.P. Stevens.

The Amalgamated Clothing and Textile Workers are also focusing on New York Life Insurance Company which manages about $1.3 billion in pension assets. Its chairman, Ralph Manning Brown, Jr., sits on the J.P. Stevens board. In their continuing use of economic clout to fight against J.P. Stevens, the participating unions intend to threaten withdrawal of funds from other institutional investors that have ties to the company.

At the end of 1977, the AFL-CIO convention evidenced a strong interest in pension assets and investment practices by adopting a key convention resolution proposing that "union funds be entrusted to financial institutions whose investment policies are not inimical to the welfare of working people, and that a portion of union funds be invested in national programs that meet public needs and provide construction jobs, such as houses, schools, hospitals, factories, and stores."[35] A general review of investment portfolios for antiunion practices was urged. The AFL-CIO is also considering the possibility of a union-label investment file.

Throughout the country there has been a new and growing recognition of pension funds as a significant political and economic tool. The construction industry pension funds recently have contributed a share of their assets (about $5 million) to a Union Labor Life Insurance Company that can only invest in construction projects using union labor.[36] Although the policy has not been implemented yet, the National Union of Hospital and Health Care Employees pension fund has initiated a resolution urging that a portion of their $236 million in assets be utilized to support socially useful programs.[37] Similarly, the Plumbing-Heating and Piping Industry of Southern California benefit fund, the largest plumbing pension fund with assets of over $181 million, has agreed to invest its resources in new construction projects thus enhancing employment opportunities for its members.[38] And for the first time since the 1940s, the United Auto Workers has attempted to influence the managers of their pension funds by urging them not to agree to the United Technologies' tender offer for Carrier Corporation stock due to the former's antilabor record. In addition, Lloyd McBride, president of the Steelworkers Union, has suggested that control of pension fund investments may become a critical question in collective bargaining with the steel industry in the near future.

The Teachers Insurance and Annuity Association-College Retirement Equities Fund (TIAA-CREF), which is one of the top identified stockholders in twenty-four corporations and one of the most powerful institutional investors, often supports corporate minority resolutions. In 1978 TIAA held $6 billion and CREF $5.9 billion in worker assets. According to one source, during 1977 "the funds supported 21 of 54 social responsibility proposals" including "asking Procter and Gamble Company to quit sponsoring 'excessively violent' television programs."[39] The investment committee, which does not buy securities in liquor or tobacco industries due to pressure from some of its affiliated colleges and universities,[40] recently urged companies included in its portfolio to cease investments in South Africa. Of course, TIAA-CREF has not had the interest in or shown its full potential for comprehensive and significant changes in corporate policies.

Attempts to utilize funds for social goals have been made by some public systems. In Hawaii and Puerto Rico, public employees may secure home mortgages from trust assets and at a lower rate than commercial loans. Nearly 44 percent of the $900 million Hawaii Retirement System is invested in mortgages. Despite bargain rates and liberal downpayments, the funds not only have remained viable but have also maintained a good rate of return.

According to Robert Tilove, an observer of pension issues, the Los Angeles City Board of Pension Commissions, in 1971, attempted to use its stock ownership in both a drug and oil company in order to influence corporate policies. San Francisco Trustees reviewed its portfolio in an attempt to discover if the trust held stock in companies guilty of pollution.[41] Presently, the California legislature is considering a bill that would mandate the establishment of social-responsibility criteria for investment of its $34 billion in state pension funds. In Rhode Island the state treasurer has indicated an interest in utilizing a portion of the state's $300 million retirement fund for middle-income housing. Other states, including Massachusetts and Wisconsin, and some cities such as Hartford and Sacramento are also reviewing investment policies.

Conclusion

With nearly $470 billion in assets, and holdings of approximately 20 percent of total corporate equities and 40 percent of corporate bonds, private as well as state and local retirement trusts have become a significant and growing source of capital in the American economy. Although organized labor has accepted pensions in lieu of wages, employees do not appear to be the main beneficiaries of the trusts. Despite this, the growing assets have the potential to serve the public interest and the needs of workers. Threat of or actual withdrawal of pension assets from selected managers and corporations as well as utilization of shareholder voting rights to influence corporate policies can be potent weapons for public- and private-sector employees.

Pressures can be exerted for research and development in product safety and industrial health, improved or expanded retirement homes for the elderly, increased pollution and other environmental controls, prounion policies, and low-interest loans or mortgages for workers. The latter would be particularly beneficial for those working-class neighborhoods in which banks have consistently refused mortgage or property improvement loans as part of a policy known as "redlining"—a geographical discrimination in mortgage lending that denies loans to individuals no matter how credit-worthy the applicant or how sound the building, and consequently contributes to urban decay.

To further these and other goals, workers will have to seek control over their retirement trusts as a key demand during collective bargaining, regain participation in decision making over those funds that have been turned over to banks and other investment managers, and urge Congress to liberalize selective investment restrictions incorporated in ERISA. Hopefully, new awareness of trust potentialities by unions and public leaders will encourage them to seek alternative means for using pension assets in the interest of workers and the communities in which they live.

Notes

1. U.S., Congress, Senate Committee on Governmental Affairs, Subcommittee on Reports, Accounting and Finances, *Voting Rights in Major Corporations* (Washington: U.S. Government Printing Office, January 1978).

2. David M. Kotz, "Finance Capital and Corporate Control," in Richard C. Edwards, Michael Reich, and Thomas E. Weisskopf, eds., *The Capitalist System* (Englewood Cliffs, N.J.: Prentice Hall, 1978), p. 156.

3. Ibid.

4. There are, however, a few recent publications that are delving into these issues. For an insightful analysis on worker pension funds and their role in the economy see Jeremy Rifkin and Randy Barber, *The North Will Rise Again* (Boston, Mass.: Beacon Press, 1978); Public Pension Funds are discussed comprehensively in Lee Webb and William Schweke, eds., *Public Employee Pension Funds: New Strategies for Investment* (Washington, D.C.: Conference on Alternative State and Local Policies, July 1979); for an interesting though less convincing argument on workers and pension fund ownership see Peter F. Drucker, *The Unseen Revolution: How Pension Fund Socialism Came to America* (New York: Harper and Row, 1976).

5. The only other expression of national policy on private plans is the Internal Revenue Act. It has been suggested that ERISA's prudent-man rule has encouraged more conservative investment practices. For example, despite new rulings by the Labor Department which attempted to clarify what is meant by prudent investments there appears to be less willingness now to invest in venture capital funds or small businesses.

6. Discussions prior to passage of ERISA focused primarily on vesting, funding requirements, insurance, portability of benefits, enhanced fiduciary standards, and liability of fund managers.

7. Roger F. Murray, *Economic Aspects of Pensions: A Summary Report* (New York: Columbia University Press, National Bureau of Economic Research, no. 85, 1968), p. 125.

8. Geoffrey N. Calvert, "Contrasting Economic Impact of OASDI and Private Pension Plans," in Richard Irwin, ed., *Social Security and Private Pension Plans: Competitive or Complementary* (Homewood, Ill.: Irwin-Dorsey, 1976), p. 68.

9. Quoted in A.H. Raskin, "Labor and Investment of Pension Funds," *The New York Times,* November 9, 1977, p. 65.

10. U.S., Congress, House Committee on Education and Labor, *Pension Task Force Report on Public Employee Retirement Systems,* 95th Cong., 2d sess. (Washington: U.S. Government Printing Office, March 15, 1978), pp. 206-207.

11. See *Pensions and Investments,* November 6, 1978, p. 31.

12. John Harrington, "New Pension Funds Investments: An Evaluation of Potential Allies and Opposition," in Webb and Schweke, *Public Employee Pension Funds,* p. 123.

13. *Pensions and Investments,* April 23, 1979, p. 1.

14. Ibid.; *Pensions and Investments* April 24, 1978.

15. Ibid.

16. Quoted in Raskin, "Pension Funds," p. 65.

17. U.S., *Congressional Record,* Proceedings and Debates of the 95th Congress, 1st session, Senator Lee Metcalf, "A Stronger Voice for Union Members in Pension Fund Investment," vol. 123, no. 198 (Washington: U.S. Government Printing Office, January 4, 1978), pp. S1 9951-9952.

18. Corporate Data Exchange, Inc., *Pension Investments: A Social Audit* (New York, 1979), p. 15.

19. Ibid.

20. See, for example, William C. Greenough and Francis P. King, *Pension Plans and Public Policy* (New York: Columbia University Press, 1976), p. 146.

21. Rifkin and Barber, *North Will Rise Again.*

22. Greenough and King, *Pension Plans,* pp. 45-46.

23. Raskin, "Pension Funds," p. 65.

24. "While Labor Looks over ERISA Retroactively," *Pensions and Investments,* October 9, 1978, p. 3.

25. *Pension Task Force Report,* pp. 87-92; presently Congress is considering a new law entitled the Public Employee Income Security Act (PERISA), modeled after the private pension act ERISA.

26. Greenwich Research Associates, Inc., "Sixth Annual Report to Executives" (Greenwich, Conn., 1978), p. 111.

27. Cited in James H. Schulz, *The Economics of Aging* (Belmont, Calif.: Wadsworth, 1976), p. 159.

28. *Pension Task Force,* p. 126.

29. Jack L. Treynor, Patrick Regan, and William Priest, Jr., *The Financial Reality of Pension Funding under ERISA* (Homewood, Ill.: Dow Jones-Irwin, 1976), p. 7.

30. Robert M. Ball, *Social Security Today and Tomorrow* (New York: Columbia University Press, 1978), p. 390.

31. See Anne M. Menzies, "The Netherlands," and Dorothy F. Wilson, "Sweden," in Thomas Wilson, ed., *Pensions, Inflation, and Growth* (London: Heinemann Educational Books, 1974), pp. 110-154, 155-200.

32. Wilson, "Sweden," p. 191-192.

33. See Wilson for further details, "Sweden," pp. 155-200.

34. H. Robert Bartell, Jr., "Growth in Multiemployer and Union Pension Funds, 1959-1964," in H. Robert Bartell, Jr., and Elizabeth T. Simpson, eds., *Pension Funds of Multiemployer Industrial Groups, Unions, and Nonprofit Organizations,* occasional paper 105, National Bureau of Economic Research (New York: Columbia University Press, 1968), pp. 1-19; Murray, *Economic Aspects,* p. 84.

35. U.S., *Congressional Record,* "A Stronger Voice," pp. S1 9951-9952.

36. "Construction Unions Invest to Aid Industry," *Pensions and Investments,* July 17, 1978, p. 44.

37. *Pensions and Investments,* June 5, 1978, p. 35.

38. "Plumbers Out of Equities," *Pensions and Investments,* November 20, 1978, p. 52.

39. *The New York Times,* March 19, 1978, p. F13.

40. Robert Tilove, *Public Employee Pension Funds* (New York: Columbia University Press, 1976), p. 218.

41. Ibid., pp. 218-219.

13 The Determinants of State Workmen's Compensation Laws

Joel A. Thompson

Introduction

The contemporary issue of workmen's compensation in the American states exhibits two noticeable characteristics: it is an immensely important issue because of its potential impact on the lives of millions of workers and yet, there is a paucity of systematic, comparative research concerning its formation and implementation.

The significance of workmen's compensation is undeniable. Since 1972 about one in ten working Americans "experienced a job-related nonfatal injury or illness or was killed because of hazards in the work environment."[1] As the United States has become more industrialized, new products and processes have been introduced into the workplace. With increasing frequency these materials have adversely affected those most intimately involved with their manufacture or production. Our recent experiences with polyvinyl chloride, kepone, asbestos, black lung, and other hazards of the workplace serve as a sad reminder that the contemporary American worker faces a different work environment than his predecessor. Since those most frequently affected by work injuries and illnesses are the ones who can least afford their debilitating effects,[2] an efficient and effective workmen's compensation program is a necessity to ensure the well-being of the labor force. However, in recent years many experts have raised serious questions about the adequacy of state workmen's compensation programs.

But despite its importance, workmen's compensation has not received a great deal of attention from political scientists and program evaluators. Indeed as one writer notes: "One of the most remarkable facts about the oldest of our social security programs is the lack of information about it."[3]

The purpose of this research is twofold: (1) to shed some light on one of the most important issues facing the American states by examining recent developments in state workmen's compensation programs and (2) to investigate the factors associated with these developments while, hopefully, avoiding some of the problems endemic to previous state policy studies.[4]

State Policy Studies

Despite the contributions of state policy studies to our understanding of variations in state policies, most studies have been plagued by simplistic assumptions and myopic conceptualizations. For example, many studies concentrated on the broad socioeconomic and political determinants of policy outputs, that is, expenditures for certain policy areas. Implicit in these studies is the assumption that these expenditures have met some intended goal, though this linkage is rarely if ever investigated.

Also in their conceptualization of political determinants, previous studies have consistently relied upon party competition or malapportionment as their primary political indicator. While the potential importance of party competition has been established, it should be recognized that party competition is not the only politically relevant variable in most states. Indeed the singular use of party competition in most studies of state policy may well be a significant reason that politics has usually been found to be of secondary importance in explaining policy variations.

Finally, most state policy studies have taken an input/output focus that ignores the processes by which states initiate, adopt, and implement policies. Thus it is impossible to ascertain *why* and *how* states adopt and implement various policies. Therefore a study that identifies and investigates the activities of important actors in the policy process can contribute to our knowledge of state policy making.

Historical Perspective of Workmen's Compensation

To fully understand the forces that have produced changes in workmen's compensation one must examine its evolution. As two scholars of workmen's compensation have stated: "workmen's compensation was not invented; it evolved. It developed out of a series of social adjustments to meet a social need."[5]

Before the principle of workmen's compensation that fixed liability upon the employer without regard to fault was adopted, a worker's only means to obtain remedy for an industrial accident and the subsequent wage loss was through the judicial system. Employer liability cases were based on the common-law precedents of negligence and tort liability. Basically, the precedents held that "occupational injuries were always the result of someone's fault, and that he should bear the costs."[6] Thus it was the responsibility of the court to determine who had been at fault and therefore responsible for the injury.

Under this procedure the "burden of proof for establishing the employer's negligence fell upon the worker, a burden often difficult, if not impossible, to fulfill."[7] Contributing to the difficulty of the injured employee in this situation was the reluctance of fellow employees to testify against their employer for fear of losing their jobs and, most importantly, the "unholy trinity of defenses" available to the employer.

Under common law there were three defenses available to the employer in an occupational injury case which composed the unholy trinity: (1) the fellow-servant doctrine, (2) assumption of risk, and (3) contributory negligence.

The fellow-servant doctrine held that the master was not responsible for the acts of his servants. This precedent was extended to the employer-employee relationship so that employees could not recover damages from their employers if it could be shown that an injury was a result of the negligent act of a fellow worker.[8] The assumption-of-risk doctrine held "that a servant, on accepting employment, assumed all the ordinary risks incident to his work."[9] Contributory negligence essentially stated that an injured worker could not recover damages if the worker had been negligent in any way, regardless of the extent to which the employer had been negligent.

However, as litigation increased with rising accident rates due to the increased mechanization and chemicalization[10] of industry in the late-nineteenth century, these defenses "soon became so riddled with exceptions and fine-spun distinctions in the master's favor that their protection to the employee was virtually nullified."[11]

The inability of common-law defenses and their later modifications to cope with the problems of industrial accidents and subsequent litigation led to the establishment of state commissions to recommend changes in state laws. These commissions "almost unanimously recommended abolition of common-law practices and employer's liability statutes in favor of a system of workmen's compensation."[12] The inadequacies of common-law practices were related to uncertain recoveries, wastefulness and high costs, delayed settlements, inconsistency of awards, the public burden created by injured workers, and the general deterioration of labor relations caused by antagonism between employer and employee in industrial accident suits.[13] Although the recommendations of the state commissions varied somewhat, "one cardinal principle was included in all the reports, namely, that liability for industrial accidents should be fixed on the employer without fault."[14]

Although the movement toward a system of workmen's compensation had been active for some time in Europe, it was not until significant changes were made in industrial processes and important groups concerned with workmen's compensation threw their support behind the concept that the workmen's compensation movement gained momentum in the United States.

In 1909 the American Federation of Labor, which had previously been opposed to workmen's compensation since it felt more could be gained for the worker under employer's liability laws, reversed its position. Later the American Bar Association became interested and studied the possibility of drawing up a uniform compensation act. Also at about this time, the National Association of Manufacturers overwhelmingly endorsed the concept of workmen's compensation.[15]

It is not surprising, then, in light of the changing economic and political environment that the next ten years witnessed a dramatic increase in workmen's compensation laws. New York was first to adopt a workmen's compensation law in 1910. In 1911, ten more states adopted workmen's compensation laws and by 1921 forty-two states had adopted workmen's compensation laws. "With labor and industry lobbying for effective compensation legislation, the movement toward reform was in full swing."[16]

While it appeared that the two decades beginning about 1908 and ending around 1928 laid the groundwork for vast improvements in workmen's compensation programs, the next four decades witnessed relatively little activity in this area. As two researchers of workmen's compensation wrote during the period, "once the pioneer trail blazer in social insurance, workmen's compensation has, like many other types of social legislation, not proved adaptable enough to keep abreast of a changing environment."[17]

Most of the initiative in workmen's compensation in the following three decades was taken by the federal government, perhaps on cue from the states. On March 4, 1927, Congress passed the Longshoremen's and Harbor Workers' Compensation Act which extended workmen's compensation benefits to employees of private employers who are engaged in maritime employment in U.S. waters. It was extended to private employees in the District of Columbia in 1928, to defense contractors in 1942, and to other workers who are employed by the federal government and its contractors in 1953 and 1958.[18]

In the 1950s the Labor Department became concerned with the provisions (or lack of them) in some state laws. It appointed a special study commission which drew up and published a *Model Act* in 1955. However, because of pressure from Congress, which viewed this as an attempt to federalize workmen's compensation, the Labor Department was forced to drop the project.[19]

Interest in workmen's compensation was renewed in the 1960s when the Council of State Governments picked up on the model act idea of the Labor Department. In 1960 the Council appointed an advisory commission which published its own recommended state legislation in 1965. The recommended standards of the council very closely coincide with those of the National Commission on State Workmen's Compensation Laws of 1972.[20]

The states were not totally inactive during this period, but the momentum of the turn of the century certainly receded. Between 1930 and 1948 the six remaining states adopted workmen's compensation laws. Several states also made some changes in their laws (primarily to include more workers) and there were other minor improvements in some states. However, while workmen's compensation "contributed significantly toward meeting the needs of the worker," its development again did not keep "pace with the social and economic advances of the American people."[21]

The relatively poor state of workmen's compensation laws in the early 1970s has been well documented by Rolland Martin's scathing critique.[22] Martin cited inequities in four areas of workmen's compensation laws:[23]

Uncoverage. In 1972 "(t)hirty percent, and more, of the employees in fifteen states (were) not covered by workmen's compensation insurance." Also many states still had elective coverage, numerical exceptions, and excluded certain hazardous industries. Nine states still limited occupational diseases to those specified in the workmen's compensation act.

Payments. "As of January 1972, more than one-half of the states paid maximum weekly benefits that (were) below the poverty level . . ." Many states still limited the dollar amount and duration of both permanent and temporary disabilities.

Medical Care. Many states still had limits on the dollar amount, duration, and type of medical services (osteopaths, nurses, occupational therapists) that an injured worker could receive.

Rehabilitation Benefits. States still had limitations on dollar amount and the duration of care. Also referral to rehabilitation centers took as long as eighteen months in some states.

"That our workmen's compensation laws (were) not meeting the needs of the worker (was) the force behind the call for reform by the National Commission on State Workmen's Compensation Laws."[24]

Contemporary Issue of Workmen's Compensation

The Occupational Safety and Health Act of 1970 established the National Commission on State Workmen's Compensation Laws. The National Commission was directed to "undertake a comprehensive study and evaluation of State workmen's compensation laws in order to determine if such laws provide an adequate, prompt, and equitable system of compensation."[25] The commission submitted its report to the president on July 31, 1972, and "criticized many aspects of state workmen's compensation programs" which basically coincided with the criticisms leveled by Martin.[26] The commission "recommended eighty-four specific recommendations for a modern workmen's compensation program"; nineteen of these were designated as essential.[27]

Each of these nineteen essential recommendations is related to the attainment of one or more of the five major objectives of a modern workmen's compensation program. These objectives and their related components will be briefly discussed.[28] The nineteen essential recommendations are listed in Appendix 13A.

1. Broad coverage of employees and of work-related injuries and diseases: Workmen's compensation laws may be classified as compulsory or elective. Under a compulsory act every employer subject to it is required to comply with its provisions for compensation of work injuries. Under an elective act an employer may accept or reject the act, but if he rejects it, he loses the customary common-law defenses of assumption of risk, contributory negligence, and the fellow-servant doctrine.

No law covers all employees. Some laws restrict coverage to hazardous industries, farm workers, domestic workers, or casual workers (occupational exemptions). Also some states cover only accidental injuries and do not include occupational diseases.

2. Substantial protection against interruption of income: There are basically five categories of injuries for which benefits are paid: (1) temporary total disability, (2) temporary partial disability, (3) permanent partial disability, (4) permanent total disability, and (5) death benefits. The benefits paid for each of the disabling injuries are usually tied to a fixed percentage of income (usually the states' average weekly wage). This percentage varies from about 50 to 100 percent. For partial disabilities, the amount recovered is usually conditional upon the injured bodily member: eye, hand, leg, and so forth. Many states also place limits on the duration of these benefits.

The purpose of death benefits is to provide income for families or other dependents of the deceased worker. These payments can be in lump sum or payable over a period of time and tied to a percentage of income. In most states, provision is also made for payment of burial allowances.

3. Provision of sufficient medical care and rehabilitation services: When a worker is injured he first needs medical aid and sometimes hospitalization. All compensation acts require medical benefits to be paid, but the amount and duration of such payment varies under different state laws. Also some states limit the kind of medical service that will be paid under workmen's compensation.

Rehabilitation of a handicapped worker attempts to restore the worker to the maximum extent in all phases of life: physical, mental, social, vocational, and economic usefulness. State laws may contain provisions for rehabilitation in the form of retraining, education, job guidance, and placement, to help the injured person make adjustments and find suitable employment.

4. The encouragement of safety: "The commissions proposing the first compensation laws in this country were of the opinion that the compensation system would materially reduce industrial accidents. Whether this hope has been realized or not, the reporting of and the compensation for industrial accidents has . . . a close relation to . . . the prevention of accidents."[29] None of the nineteen essential recommendations is specifically concerned with this objective and the Occupational Health and Safety Administration is now responsible for worker safety (in conjunction with some state agencies). Since this overlapping jurisdiction makes analysis impossible, it will not be examined.

5. An effective system for the delivery of benefits and services: "The fifth objective recommended by the National Commission is a means to an end, not an end in itself. Its performance is evaluated relatively in comparison with other systems and absolutely by the degree of accomplishment of the four basic objectives of workmen's compensation."[30] Again since none of the nineteen essential recommendations specifically concerns this objective it will not be discussed at length here. However, previous research suggests that the other objectives are to

some extent dependent on system characteristics.[31] Thus the following discussion of workmen's compensation will be limited to compliance with the nineteen essential recommendations that are directed toward the attainment of the first three objectives: broad coverage, income protection, and medical and rehabilitation services.

Efforts toward Compliance

Table 13-1 lists the nineteen essential recommendations along with state compliance totals for 1972 and 1977, and a corresponding change in compliance measure. (Recommendation 2.1 has two components and recommendation 3.25, four components. Each of these components was treated as a recommendation making the maximum total compliance score 23.)

One of the first observations made from table 13-1 is the general low levels of compliance in 1972. Forty-two states were in compliance with less than half of the recommendations in 1972 and some states were in compliance with only three or four recommendations. These observations substantiate the findings of both Martin and the National Commission concerning the inadequacies of state workmen's compensation programs.

However, between 1972 and July 1977 most states made significant gains in compliance with the recommendations of the National Commission. By 1977 thirty-three states were in compliance with more than half of the recommendations while only a few states had made no progress or had changed their laws in such a manner that they were no longer in compliance with the same number of recommendations as in 1972. Overall there was about 34 percent increase in compliance between 1972 and 1977. However, as table 13-1 illustrates, changes in compliance were uneven. Some states changed rather dramatically (Montana and Louisiana) while others changed very little or none despite low levels of compliance in 1972 (Florida, Mississippi, and Oklahoma). An interesting question to be investigated then is: what factors have a significant impact on compliance with workmen's compensation recommendations?

Factors Associated with Compliance

In selecting variables as potential contributors to an explanation of workmen's compensation legislation, two sources were utilized. First, factors that have historically been related to changes in workmen's compensation laws are obviously important. Second, since this work also attempts to circumvent some weaknesses of previous state policy studies, it is necessary that variables that have been shown to be important determinants of other state policies be included in the analysis for comparability. Since workmen's compensation is a state policy, it is not surprising that the same variables emerge from both sources.

Table 13-1
State Workmen's Compensation Laws Compared with the Nineteen Essential Recommendations of the National Commission on State Workers' Compensation Laws

	Essential Recommendation																						State Compliance Score	State Compliance Score	Change
	2.1																*3.25*								
State	*(a)*	*(b)*	*2.2*	*2.4*	*2.5*	*2.6*	*2.7*	*2.11*	*2.13*	*3.7*	*3.8*	*3.11*	*3.12*	*3.15*	*3.17*	*3.21*	*3.23*	*(a)*	*(b)*	*(c)*	*(d)*	*4.2*	*4.4*	*1977*	*1972*	*1977–1972*
AL	X	X	–	–	–	–	X	X	X	X	–	X	X	–	X	–	–	–	–	–	–	X	X	11	2	9
AK	X	X	X	–	–	–	–	X	X	X	X	X	X	X	X	X	X	–	–	–	–	X	X	15	10	5
AZ	X	–	X	X	–	X	X	X	X	X	–	X	X	–	X	–	–	–	–	–	–	X	X	13	9	4
AR	X	X	–	–	–	X	–	–	X	X	–	X	X	–	X	–	–	–	–	–	–	X	X	10	5	5
CA	X	X	X	X	–	X	X	–	X	X	–	X	X	–	–	X	–	–	–	–	–	X	–	12	8	4
CO	X	–	X	(X)	–	X	–	X	X	X	–	X	X	–	X	X	–	–	–	–	–	X	X	13	10	3
CT	X	–	X	X	–	–	–	–	X	X	–	X	X	–	X	X	–	–	–	–	–	X	X	11	14	-3
DE	X	X	X	–	–	–	–	X	X	X	–	X	X	–	X	X	–	–	–	–	–	X	X	12	9	3
FL	X	–	X	–	–	X	–	X	X	–	–	X	–	–	X	–	–	–	–	–	–	X		8	6	2
GA	X	–	–	–	–	–	–	X	X	X	–	X	X	–	X	X	–	–	–	–	–	X	X	10	6	4
HI	X	X	X	X	–	X	–	X	X	X	X	X	X	X	X	–	–	X	X	X	–	X	X	18	18	0
ID	X	X	X	–	–	X	–	X	X	–	–	X	–	–	X	–	–	–	–	–	–	X	X	10	10	0
IL	X	X	X	–	–	–	–	X	X	X	X	X	X	X	X	X	X	–	–	–	–	X	X	15	6	9
IN	X	X	X	–	–	X	–	X	X	X	–	X	X	–	–	X	–	–	–	–	–	X	X	12	7	5
IA	X	–	X	–	–	X	–	X	X	X	X	X	X	X	X	X	X	–	–	–	–	X	X	15	10	5
KS	X	–	–	–	–	X	X	X	X	X	–	X	X	–	–	X	–	–	–	–	–	X	–	10	1	9
KY	X	–	X	–	–	X	X	X	X	X	–	X	X	–	X	–	–	–	–	–	–	X	X	12	8	4
LA	X	X	X	X	–	–	–	X	X	X	–	X	X	–	X	–	–	X	X	X	–	X	X	15	2	13
ME	X	–	X	–	–	X	X	–	X	X	X	X	X	X	X	X	X	–	–	–	–	X	–	14	12	2
MD	X	–	X	–	–	X	X	–	X	X	X	X	X	X	X	X	X	X	X	X	–	X	X	18	18	0
MA	X	X	X	X	–	–	–	–	X	X	–	X	X	–	–	–	–	–	–	X	X	X	X	12	10	2
MI	X	X	X	X	–	X	X	–	X	X	–	–	X	–	X	X	–	–	–	–	–	X	X	13	14	-1
MN	X	X	X	–	–	–	X	–	X	X	X	X	X	X	X	–	X	X	–	X	–	X	X	16	11	5
MS	X	X	–	–	–	–	–	–	X	X	–	X	X	–	–	–	–	–	–	–	–	X	X	8	8	0

MO	X	X	–	–	–	X	X	X	X	X	–	X	X	–	X	X	–	X	X	X	–	X	–	15	10	5
MT	X	X	X	X	–	X	–	–	X	X	X	X	X	X	X	X	X	X	X	X	X	X	X	20	4	16
NE	X	X	X	–	–	X	X	X	X	X	–	X	X	–	X	X	–	X	X	X	X	X	X	18	11	7
NV	X	X	X	–	–	X	–	–	X	X	X	X	X	X	X	X	X	–	–	–	–	X	X	15	9	6
NH	X	X	X	X	X	X	X	X	X	X	X	X	X	X	X	X	X	–	–	X	X	X	X	21	14	7
NJ	–	X	X	X	–	X	X	X	X	X	–	X	X	–	–	–	–	–	–	–	–	X	–	11	11	0
NM	X	X	–	–	–	X	–	X	X	X	X	X	X	X	–	–	X	–	–	–	–	X	X	13	2	11
NY	X	X	X	–	–	–	–	–	X	X	–	X	X	–	X	–	–	–	–	–	–	X	X	10	10	0
NC	X	–	–	–	–	–	–	X	X	X	X	X	X	X	X	X	X	–	–	–	–	X	X	13	3	10
ND	X	X	X	–	–	X	–	–	X	X	X	X	X	X	X	X	–	X	–	X	–	X	X	16	11	5
OH	X	–	X	X	–	X	X	–	X	X	X	X	X	X	X	X	X	X	X	X	X	X	X	20	10	10
OK	X	–	–	–	–	–	–	–	–	X	–	X	X	–	–	–	–	–	–	–	–	X	X	6	7	-1
OR	X	X	X	X	–	X	–	–	X	X	X	X	X	X	X	–	–	X	–	X	–	X	X	16	13	3
PA	X	X	X	–	–	–	–	X	X	X	X	X	X	X	X	–	X	–	–	–	–	X	X	14	11	3
RI	X	X	–	–	–	–	X	–	X	X	X	X	X	X	X	X	X	X	–	X	–	X	X	16	12	4
SC	–	–	–	–	–	–	–	X	X	X	X	X	X	X	–	X	X	–	–	–	–	X	X	11	4	7
SD	X	–	X	–	–	–	–	–	X	X	–	X	X	–	X	X	–	–	–	–	–	X	X	10	7	3
TN	X	–	–	–	–	–	X	–	X	X	–	X	X	–	–	–	–	–	–	–	–	X	X	8	2	6
TX	–	X	X	–	–	–	–	–	X	X	–	X	X	–	–	X	–	X	X	X	X	X	X	13	5	8
UT	X	X	X	–	–	X	X	–	X	X	X	X	X	–	X	X	–	–	–	–	–	X	X	14	10	4
VT	X	–	X	–	–	–	X	X	X	X	X	X	X	X	–	X	X	–	–	–	–	–	–	12	7	5
VA	X	–	–	–	–	–	–	–	X	X	X	X	X	X	–	X	X	–	–	–	–	X	X	11	8	3
WA	X	X	X	–	–	X	X	X	X	–	–	X	–	–	X	–	–	–	–	–	–	X	X	11	15	-4
WV	X	X	X	–	–	X	X	–	X	X	X	X	X	X	–	X	X	X	–	X	–	X	X	17	8	9
WI	X	X	X	–	–	X	X	X	X	X	X	X	X	X	X	X	X	–	–	–	–	X	X	17	11	6
WY	X	X	X	–	–	–	–	X	X	X	X	X	–	–	–	–	–	–	–	–	–	X	X	10	8	2
Totals	48	33	38	13	1	29	22	27	50	48	25	50	47	23	36	30	20	13	8	15	6	50	44	661	437	33.9%

Source: U.S. Department of Labor.

Laws in Effect on July 1, 1977.

X Law meets recommended standard.

– Law does not meet recommended standard.

(X) Law has been enacted which will bring state into compliance at a future date, not later than June 30, 1978.

Socioeconomic Variables. Socioeconomic variables such as levels of income, urbanization, and education have been shown to be important determinants of some types of policy outputs.[32] As Thomas Dye noted, these variables are usually employed as indicators of economic development.[33]

Theoretically it is expected that a relationship between workmen's compensation and economic development exists. As states become more economically developed, larger portions of the work force are drawn into the industrial sector. The increasing mechanization and chemicalization of this industrial activity in turn leads to a more hazardous work environment, more industrial accidents and illnesses, and subsequently for greater demands for more liberal workmen's compensation benefits and services.[34] As measures of economic development, this research will employ (1) median family income, 1969; (2) percentage of urbanization, 1970; and (3) median school years completed, 1970.

Political Variables. Political variables, primarily party competition, have also been shown to be important determinants of some types of policy outputs.[35] However, political characteristics have usually taken a backseat in most policy studies to socioeconomic factors. As suggested in the introduction, this may in part be due to the absence of other *relevant* political factors rather than necessarily the lack of importance of politics in general. With this in mind, two somewhat different measures of political characteristics will be included in this analysis: interest group involvement and party competition. As Grant McConnell noted:[36]

> In state government and politics, there are many opportunities for private influence over or control of public policies. The small constituencies of states and the serious fragmentation of their political systems give functional groups—whether trade, farm, or professional—great advantages which they have nowhere been reluctant to seize.

McConnell goes on to note that interest groups are active at the state level because they are protected by state constitutions and the "political disorganization" of state governments in general.[37] Who then benefits from this disorganization? "The advantages of disorganized politics accrue quite impartially to whatever groups, interests, or individuals are powerful in any way."[38] And at the state level Zeigler and Peak noted that groups such as labor unions and trade and manufacturers associations are "very powerful interest groups in contemporary America."[39]

Not only are labor and business groups important for a general study of public policy in the American states, but they are especially important for a study of workmen's compensation. As noted previously, these two groups historically have been involved in workmen's compensation. Also it can be argued (and several contemporary case studies support the contention) that workmen's compensation is a policy with labor/management dimensions.[40] Two other groups which have also been involved in workmen's compensation are insurance and the legal profession.[41]

To measure interest-group involvement in workmen's compensation, an index of involvement for each group was constructed. Information used in constructing the index came from a mail survey of members of legislative committees that deal with workmen's compensation legislation in each house of each state legislature.[42]

The final political variable utilized is party competition. While several studies of workmen's compensation have stressed the involvement of interest groups in this policy, none has suggested that party organizations are very active. This situation underscores the importance of identifying the relevant political variables that interact in any policy area. Since party competition has received so much attention in the literature of state policy it will be included in this analysis, though its importance is expected to be less than that of interest-group involvement.[43]

The relationship between these two political characteristics has been given some attention in the literature of political science. Perhaps the two best studies at the state level, though not comprehensive, are those of Harmon Zeigler[44] and Grant McConnell.[45] Both studies reach similar conclusions: (1) interest groups are stronger where parties are weak, (2) interest groups are stronger in one-party states, and (3) interest groups are stronger in states that lack diversity of interests. Thus there appears to be a reciprocal relationship between interest group and party strength which emphasizes the importance of examining both when considering political characteristics.

Findings

Table 13-2 presents the bivariate correlations between compliance with workmen's compensation recommendations and the selected socioeconomic and political indicators. Emerging from table 13-2 is a relatively balanced picture of factors associated with compliance. Both socioeconomic and political factors show a strong relationship with compliance, and insurance involvement is the only variable associated with change in compliance between 1972 and 1976. However, since these variables are interrelated, controls must be imposed before more definitive conclusions can be drawn.

Table 13-3 presents the beta coefficients for each of the selected variables which allows for a comparison of the relative impact of each variable on compliance with workmen's compensation recommendations. Again the findings reveal that workmen's compensation legislation appears to be a function of both socioeconomic and political influences within a state. Wealth, as usual, is found to be an important determinant of state policy but so are the activities of several vitally concerned interest groups.[46] These groups have variously been

Table 13-2
Correlations between Compliance and Selected Socioeconomic and Political Indicators

Variables	*Compliance 1977*	*Change in Compliance 1977-1972*
Socioeconomic		
Income	.50	.05
Urbanization	.18	.01
Education	.38	.07
Political		
Labor involvement index	.39	.13
Business involvement index	.14	.13
Insurance involvement index	.38	.31
Legal involvement index	.20	.10
Party competition	.07	.06

Notes: Since the universe of cases is used, no significance levels are included.
Change in compliance was computed from:

$$\text{Change in compliance} = \frac{\text{compliance score 1977 - compliance score 1972}}{\text{23 - compliance score 1972}}$$

Table 13-3
The Determinants of Workmen's Compensation Legislation

Variables	*Compliance*[a] *1977*	*Change in Compliance*[a] *1977-1972*
Socioeconomic		
Income	.39	-.20
Urbanization	-.23	.12
Education	.16	.11
Political		
Labor involvement index	.27	.11
Business involvement index	-.06	.05
Insurance involvement index	-.21	-.42
Legal involvement index	.09	29
Party competition	-.07	.08

[a]Beta coefficients.

cited as the prime movers and prime obstructors of workmen's compensation reform in the states.

For example, a study of workmen's compensation reform in Texas led one researcher to assert that the process was "best understood in the interaction of the workmen's compensation interest groups within the Texas legislature." He concluded that the workmen's compensation system in Texas helped "to maintain most of the interest organizations and institutions in this field" and was "probably the single most influential force in developing and maintaining the Texas Manufacturers Association as an institution."[47] Table 13-3 generally confirms these expectations. Labor involvement is positively associated with higher levels of compliance while insurance involvement is inversely associated with compliance. These findings lead to a conclusion similar to that of the National Commission: "(i)n many States, substantial reform is difficult because there is more than one interest group with power to veto proposed changes in the law, and it is difficult to find a package of amendments acceptable to all parties."[48]

It is interesting to note that party competition is not an important determinant of workmen's compensation legislation. This result is not surprising since workmen's compensation appears to be an issue that is articulated by interest groups rather than party structures within the states. However, had this research defined the political arena in terms of party politics only, the findings would have been similar to those reached by previous researchers that socioeconomic variables are the prime determinants of state policies.

Conclusions

This research traced recent developments in workmen's compensation programs. In general, it was found that since the National Commission issued its report in 1972 the states have been active in complying with the nineteen essential recommendations made by the National Commission. Indeed, given the problems faced by many state legislatures (short sessions, inadequate staff) the states have moved with commendable speed in this area.

One obviously important question that this research leaves unanswered is whether or not compliance in and of itself has significantly affected the quality of state workmen's compensation programs. In other words, does compliance with these recommendations mean that injured workers are receiving better benefits and services? In a previous work the author investigated this question and the findings led to the conclusion that, while compliance appears to have some impact on the delivery of benefits and services, other factors also play a major role.[49] This linkage between legislative outputs and actual impact deserves further investigation.

A final objective of this research was an attempt to identify and ameliorate shortcomings of previous policy studies. Hopefully, to some extent this objective has been met. Having identified the important actors in workmen's compensation in the states, this research found strong evidence that these actors are intimately involved in the articulation and promulgation of workmen's compensation policy. In this respect, evidence has been found that puts politics back into policy.

Notes

1. U.S. Department of Labor (Bureau of Labor Statistics Bulletin no. 1981) *Occupational Injuries and Illnesses in the United States by Industry, 1975* (Washington: U.S. Government Printing Office, 1975), p. 1.

2. See Daniel N. Price, "A Look at Workers' Compensation Beneficiaries," *Social Security Bulletin* 39 (October 1976) for a discussion of the personal and economic characteristics of workmen's compensation beneficiaries.

3. Herman M. Somers and Anne R. Somers, *Workmen's Compensation: Prevention, Insurance, and Rehabilitation of Occupational Disability* (New York: John Wiley, 1954), p. ix.

4. Biographies tracing the development of this research are numerous and need not be repeated here. The interested reader is referred to John H. Fenton and Donald N. Chamberlayne, "The Literature Dealing with the Relationship between Political Process, Socioeconomic Conditions and Public Policies in the American States: A Selected Bibliographic Essay," *Polity* 1 (Spring 1969): 388-404.

5. Somers and Somers, *Workmen's Compensation,* p. 17.

6. Ibid., pp. 17-18.

7. Ibid., p. 18

8. Ibid.

9. Walter F. Dodd, *Administration of Workmen's Compensation* (New York: Commonwealth Fund, 1936), p. 4.

10. Somers and Somers, *Workmen's Compensation,* p. 8.

11. Dodd, *Workmen's Compensation,* p. 8.

12. Somers and Somers, *Workmen's Compensation,* p. 22.

13. Ibid., pp. 22-26.

14. Dodd, *Workmen's Compensation,* p. 25.

15. Somers and Somers, *Workmen's Compensation,* p. 31.

16. Marcus Rosenblum, ed., *Compendium on Workmen's Compensation* (Washington: U.S. Government Printing Office, 1973), p. 17.

17. Somers and Somers, *Workmen's Compensation,* p. 269.

18. Department of Labor, *Longshoremen's and Harbor Worker's Compensation Act* (Washington: U.S. Government Printing Office, 1966).

19. Arthur Larson, "Compensation Reform in the United States," in E. Cheit and M. Gordon, eds., *Occupational Disability and Public Policy* (New York: John Wiley, 1963), pp. 29-30.

20. Ibid., p. 30.

21. Department of Labor, *Workmen's Compensation: The Administrative Organization and Cost of Administration* (Washington: U.S. Government Printing Office, 1966), p. 99.

22. Rolland Martin, *Occupational Disability: Causes, Prediction, Prevention* (Springfield, Ill.: Charles C. Thomas, 1975).

23. Ibid., pp. 14-25.

24. Ibid., p. vii.

25. *Supplemental Studies for the National Commission on State Workmen's Compensation Laws* (Washington: U.S. Government Printing Office, 1973), p. 13.

26. Council of State Governments, *Workmen's Compensation: A Challenge to the States* (Lexington, Ky.: Council of State Governments, 1973), p. ix.

27. Ibid.

28. *Summary of State Workmen's Compensation Laws,* Wage and Labor Standards Administration (Washington: Government Printing Office, 1970), pp. 1–7.

29. Dodd, *Workmen's Compensation,* pp. 59–60.

30. *The Report of the National Commission on State Workmen's Compensation Laws* (Washington: U.S. Government Printing Office, 1973), p. 99.

31. See Joel A. Thompson, "Workmen's Compensation and Public Policy in the American States" (Ph.D. diss., University of Kentucky, 1979), chs. 5 and 6 for a discussion of the requisites of an adequate workmen's compensation system and its impact on the other objectives.

32. Richard Dawson and James Robinson, "Inter-party Competition, Economic Variables, and Welfare Policies in the American States, *Journal of Politics* 25 (May 1963): 265–289; Thomas Dye, *Politics, Economics and the Public* (Chicago: Rand McNally, 1966); Phillip Roeder, *Stability and Change in the Determinants of State Expenditures* (Beverly Hills, California: Sage, 1976).

33. Thomas Dye, *Politics in States and Communities* (Englewood Cliffs, N.J.: Prentice-Hall, 1977), p. 8.

34. Somers and Somers, *Workmen's Compensation,* p. 8.

35. Charles Cnuddle and Donald McCrone, "Party Competition and Welfare Policies in the American States, *American Political Science Review* 63 (September 1969): 858–866.

36. Grant McConnell, *Private Power and American Democracy* (New York: Random House, 1966), p. 8.

37. Also see Lewis Froman, "Some Effects of Interest Group Strength in State Politics," *American Political Science Review* 60 (December 1966): 952–962.

38. McConnell, *Private Power,* p. 178.

39. Harmon Zeigler and Wayne Peak, *Interest Groups in American Society* (Englewood Cliffs, N.J.: Prentice-Hall, 1977), p. 234.

40. Tony Korioth, "The Forces That Produce Change in the Workmen's Compensation Laws of Texas and Louisiana," *Supplemental Studies for the National Commission on State Workmen's Compensation Laws* (Washington: U.S. Government Printing Office, 1973).

41. For the involvement of these groups in some states see articles by Tony Korioth, Arthur Motley, and Keith Skelton in *Supplemental Studies.*

42. This survey was part of a larger one of state legislators and state workmen's compensation administrators conducted in spring 1978. In total 161 usable responses from legislators in 47 states were received. The responses were then aggregated for each state using a weighted average as the measure of central tendency.

It should be added that for the great majority of the responses to the two questions used to construct the index an excellent consensus was found. Almost one-third (31 percent) of the responses for each state had standard deviations of zero while only one in ten had a standard deviation greater than one on a four or five point Likert scale question.

Though this measure is based on the perceptions of a relatively few individuals, they are perhaps in a better vantage point to make judgments about the strength and activity of these organizations than most.

The index was constructed from two questions. One dealt with perceptions of how much power and influence each group has within the state (4 = great deal, 3 = moderate, 2 = some, 1 = little or none). The other dealt with the intensity of feeling each group has on each component of workmen's compensation: broad coverage, increased benefits and medical payments (coded +2 = strongly favor, +1 = favor, 0 = neutral, -1 = oppose, -2 = strongly oppose). These two responses were then multiplied to obtain the index.

43. Party competition will be measured using a modification of Ranney's classification of states according to their degree of party competition. Ranney, "Parties in State Politics," in Herbert Jacob and Kenneth Vines, eds., *Politics in the American States* 3d. ed. (Boston: Little, Brown, 1971), p. 87. States with a value of less than .5000 had .5000 added to their score while states with a value greater than .5000 had .5000 subtracted from their score. This measure for each state was then subtracted from 1 to create an index of party competition.

44. Harmon Zeigler, "Interest Groups in the State," in Jacob and Vines, *Politics in the American States* (Boston: Little, Brown, 1965).

45. McConnell, *Private Power.*

46. For an excellent discussion of the activity of these groups in Louisiana and Texas, see Tony Korioth, "Forces That Produce Change."

47. Ibid., pp. 518–519.

48. *Report,* p. 119.

49. Joel A. Thompson, "The Impacts of State Workmen's Compensation Programs" (Paper presented at the Annual Meeting of the Southern Political Science Association, Atlanta, Georgia, 1978).

Appendix 13A: Essential Recommendations of the National Commission on State Workmen's Compensation Laws

R2.1 Coverage by workmen's compensation laws be compulsory and that no waivers be permitted.

R2.1(a) Coverage is compulsory for private employments generally.

R2.1(b) No waivers are permitted.

R2.2 Employers not be exempted from workmen's compensation coverage because of the number of their employees.

R2.4 A two-stage approach to the coverage of farmworkers. First, as of July 1, 1973, each agriculture employer who has an annual payroll that in total exceeds $1,000 be required to provide workmen's compensation coverage to all of his employees. As a second stage, as of July 1, 1975, farmworkers be covered on the same basis as all other employees.

R2.5 As of July 1, 1975, household workers and all casual workers be covered under workmen's compensation at least to the extent they are covered by Social Security.

R2.6 Workmen's compensation coverage be mandatory for all government employees.

R2.7 There be no exemptions for any class of employees, such as professional athletes or employees of charitable organizations.

R2.11 An employee or his survivor be given the choice of filing a workmen's compensation claim in the State where the injury or death occurred, or where the employment was principally localized, or where the employee was hired.

R2.13 All States provide full coverage for work-related diseases.

R3.7 Subject to the State's maximum weekly benefit, temporary total disability benefits be at least 66 2/3 percent of the worker's gross weekly wage.

R3.8 As of July 1, 1973, the maximum weekly benefit for temporary total disability be at least 66 2/3 percent.

R3.11 The definition of permanent total disability used in most States be retained. However, in those few States which permit the payment of permanent total disability benefits to workers who retain substantial earning capacity, the benefit proposals be applicable only to those cases which meet the test of permanent total disability used in most States.

R3.12 Subject to the State's maximum weekly benefit, permanent total disability benefits be at least 66 2/3 percent of the worker's gross weekly wage.

R3.15 As of July 1, 1973, the maximum weekly benefit for permanent total disability be at least 66 2/3 percent of the State's average weekly wage, and that as of July 1, 1975, the maximum be at least 100 percent of the State's average weekly wage.

R3.17 Total disability benefits be paid for the duration of the worker's disability, or for life, without any limitations as to dollar amount or time.

R3.21 Subject to the State's maximum weekly benefit, death benefits be at least 66 2/3 percent of the worker's gross weekly wage.

R3.23 As of July 1, 1973, the maximum weekly death benefit be at least 66 2/3 percent of the State's average weekly wage, and that as of July 1, 1975, the maximum be at least 100 percent of the State's average weekly wage.

R3.25 Death benefits be paid to a widow or widower for life or until remarriage, and in the event of remarriage two years' benefits be paid in a lump sum to the widow or widower. Benefits for a dependent child be continued at least until the child reaches 18, or beyond such age if actually dependent, or at least until age 25 if enrolled as a full-time student in any accredited educational institution.

R3.25(a) Benefits are paid to spouse for life or until remarriage.

R3.25(b) Two years' benefits paid in lump sum, in event of remarriage.

R3.25(c) Benefits paid to a child at least until age 18 and beyond if actually dependent.

R3.25(d) Benefits paid to full-time student dependents until age 25.

R4.2 There be no statutory limits of time or dollar amount for medical care or physical rehabilitation services for any work-related impairment.

R4.4 The right to medical and physical rehabilitation benefits not terminated by the mere passage of time.

Source: *Report of the National Commission on State Workmen's Compensation Laws*, Washington: U.S. Government Printing Office, 1972.

Part V
Public Employee Unions

14 The Future of Local Public-Sector Unionism: Adaptation to New Circumstances

James H. Seroka

Introduction

The establishment of reliable and valid trends in the field of local public employee unionism is a very complex and difficult task and it inherently suffers from two major weaknesses. One weakness is the inevitable loss of validity that results from the application of a similar set of criteria to a wide variety of disparate personnel functions. Supervisory personnel, hospital workers, sanitation workers, teachers, social workers, and so on, are all local public service employees; yet the dynamics of their group growth and their public-sector bargaining positions can vary enormously. Overlooking these differences ignores a crucial aspect of the labor-management situation in local governments.

A second weakness in general forecasting about the future of local public-sector labor-management relations is the unavoidable loss of reliability resulting from a concentration on mean trends at the expense of regional and spatial differences. In other words, the broad trends that exist in public-sector unionism may not be able to explain adequately and completely the interaction of labor and management in localities as disparate as New York City; Danville, Illinois; and Wilkes County, North Carolina. In short, this study is designed to describe the projected mean level of behavior in public-sector unionism for the next several years. It cannot and does not deal with the wealth of functional and spatial variation that exists in this field.

In the past, several scholars and labor observers made predictions about the direction and scope of collective bargaining and labor-management relations in the public sector. One group of scholars including Cohany and Dewey (1970), McCarthy (1970), Stanley (1972), and Saltzstein (1974) projected losses in the ability of administrators and elected officials to maintain complete control over vital management and personnel functions when dealing with public employee groups. For example, Stanley argued:

> Urban public adminsitration is undergoing major changes as a result of union pressures. A "whole new ball game" has started since unions in the public sector have begun to operate in areas traditionally reserved to management. While the force of the union impact varies by locality . . . and by function . . . , the central fact is change, and the rate of change is increasing. Pay, benefits and working conditions are more

> and more being determined bilaterally, and the scope of union agreements is expanding significantly.[1]

Other observers, such as Toledano (1975), are much less sanguine in their views, and they have projected that there is a major threat to American democracy and freedom from union involvement in public management functions. The general conclusions reached by all these studies, however, is that union power is growing at the expense of management.

A second group of public-sector labor commentators, not necessarily distinct from the first, have concentrated upon the political impact of public employee union growth. They have predicted an increasing dominance of public-sector unions over the local political environment (Love & Sulzner 1972). Bakke (1970) and Couturier (1970) link this concern with heightened militancy and strike activity. Wellington and Winters (1971) go somewhat further and imply that competing centers of political power in local governments will atrophy and wither from the onslaught of the powerful public employee unions.

In retrospect, two points are most noteworthy from this brief examination of earlier projections of the future impact of collective bargaining in the public sector. One is the innate hostility or fear of public-sector unionization which is shared by nearly every commentator who engaged in this exercise.[2] The second point is the extent to which events have not substantiated their claims in recent years. The volume and magnitude of local public employee strikes has not increased at the expected rate, growth in membership has leveled off, and nowhere have local governments surrendered their authority to a public-sector union.[3]

In this study I take a fresh look at the public-sector bargaining environment in order to explain the anomaly between expectations and reality, and to reassess and reformulate the predictions for the future. In particular, I construct a model of the public employee bargaining process, evaluate that model in light of recent developments, discuss the implications from the expected changes, and project the overall future trends in this area.

The Model

The bargaining position and strength of a public-sector union is not static over time; it shifts with shifts in the interaction of a wide set of variables. Labor-management relations in the public sector also do not exist in a vacuum and are dependent upon these variables as well. At a minimum we must consider the following factors: the socioeconomic and attitudinal climate, public employee union strength, political support, fiscal characteristics, and managerial independence and coherence. For convenience, this truncated set of variables and probable patterns of influence is outlined in a simple model in figure 14-1.

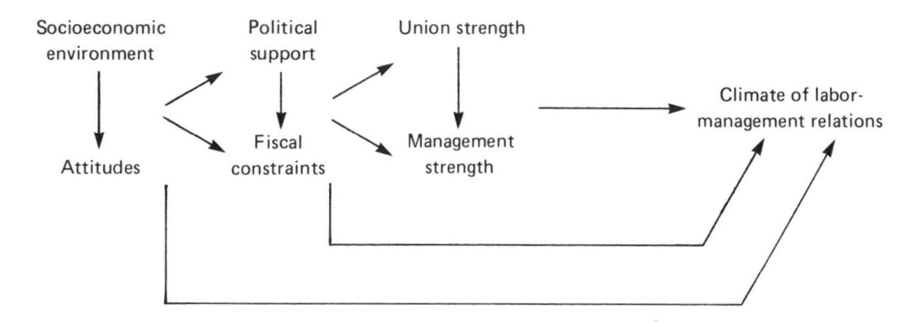

Figure 14-1. Simplified Model of the Public-Sector Collective Bargaining Environment

The model contains several useful features: it graphically illustrates the interdependence of various societal elements on the collective bargaining climate. It also highlights the necessity to place temporary limits on projections, and finally it provides us with a set of criteria for analysis of the problem.

Several variables were not included in the model. One is variation in legal controls. It is downplayed since major changes in the legal climate of public-sector bargaining have not occurred in recent years and, as Burton and Krider (1975) and Anderson (1979) discovered, because legal restraints seem to be largely irrelevant to the process. City structure is also neglected, both on the basis of empirical evidence of its relative weakness on the collective bargaining climate (Ehrenberg 1973) and on the fact that recent years have not witnessed major shifts in formal structural arrangements. In addition, the role of federal and state authorities has been limited to the fiscal arena in the model, largely resulting from a series of research findings, including Hale and Palley (1979) and Wright (1978) which indicate that federal ability to influence local behavior is limited and, except for rare exceptions (such as desegregation), is not clearly and consistently communicated to local officials. Finally, figure 14-1 does not portray the wide possible range of interactions that exist among the individual variables, though they will be considered in the analysis.

Trends

The systemic environment of municipal and county labor relations has not remained stagnant over time. Various new factors have begun to impinge upon the prevailing equilibrium supposedly prevailing in this area for the past decade. The strength of these changes is just as powerful, if not more powerful, than those that ushered in the growth of membership and power of public employee unions in the 1960s, and the impact of these new changes in the systemic environment may be quite different from those that preceded it. For convenience, I will concentrate on only three areas of structural constraints that affect local public-sector labor relations. They are the fiscal, managerial, and demographic areas.

Fiscal

The nature of fiscal constraints has long had a strong impact upon the quality of labor relations and public employee unionization in the United States. These fiscal constraints, if nothing else, can determine the parameters of flexibility present in labor-administrative relations in the local public sector. Today the local fiscal system has changed, especially in the areas of inflation, revenue limitations, and intergovernmental funding.

Inflation. Local tax revenues have been increasing at a rate faster than that of general inflation during the period 1972-1977. In addition, public employee income has not kept pace with the gains in general revenue. As indicated in table 14-1, the early 1970s were a very difficult period for local fiscal management since all the major components of local budgeting costs were engaging in an escalating inflationary race with each other.

The inflationary situation has muddied considerably the strategy, tactics, and relative position of public-sector unions and management personnel. From the union perspective, the gains received after a long and arduous strike can be erased very rapidly in an inflationary period. This situation forces a union to

Table 14-1
Annual Percentage Growth in Local Tax Revenues, Consumer Price Index, GNP Deflator, and in Average Monthly Public Employee Personnel Costs, 1972-77

Year	*Tax Revenue*	*Consumer Price Index*[a]	*GNP Deflator*[a]	*Personnel Costs*
1972-73	9.4	4.3	5.1	8.7
1973-74	5.5	3.3	4.1	6.0
1974-75	8.7	6.2	5.9	6.4
1975-76	10.5	11.0	9.7	7.2
1976-77	11.5	9.1	9.3	6.0

Sources: U.S. Department of Commerce, Bureau of the Census, *Public Employment in 1977* (1978):8; *Statistical Abstract of the U.S. 1977* (1978):471; *Taxes and Intergovernmental Revenues 1977* (1978); and *Taxes and Intergovernmental Revenues 1976* (1977).

[a]Since future tax decisions and personnel demands are based on then current price conditions, the previous years index figures are used.

choose from three possible alternatives: (1) to encourage its members to tighten their belts, thereby losing credibility and support; (2) to encourage its members to again take to the picket line; or (3) to seek a more permanent modus vivendi with the local government management in order to provide the best possible solution for its members.

From management's perspective, the inflationary spiral can provide an opportunity to reassert the initiative in collective bargaining (assuming that it had been lost) and work toward the development of a mature bargaining relationship. From another perspective, however, the inflationary spiral makes it extremely difficult to continue with incremental decision making in the development of personnel and wage policy, and therefore encourages the adoption of an innovative bargaining situation and possible conflicts.

Revenue Limitations. Tax revolt measures such as California's Proposition 13, or the threat of such measures, are beginning to place severe constraints upon the ability of a local government to bargain with local employee associations. In particular, these tax revolt referendum issues, ranging from Nevada to Michigan, have reduced flexibility in bargaining in wage and expenditure questions. Local governments can claim they lack the resources to grant equitable wage settlements and be believed. Ohio's recent experiences with school bond millage defeats in Dayton, Medina, and Cleveland clearly illustrates the potential impact of publicly imposed constraints on revenue.

If the adoption of tax cut proposals spreads beyond California, recent public employee salary and fringe benefits gains may wither away. Nevertheless, even if similar propositions are not adopted, the perception that expenditures must be cut will be widely shared and acted upon (*Business Week* 1975). Therefore, since personnel is a major budget factor, accounting for over one-half of a local government's expenditure, employee relations will be sharply affected (Benecki 1978).[4]

Intergovernmental Trends. The flexibility of local units of government in dealing with basic revenue and expenditures decisions may be reduced even further if present trends in intergovernmental finance continue. In recent years, intergovernmental aid has become the largest source of local revenue funds replacing local taxes in importance in 1975. As table 14-2 indicates, the trend has been gradual, but its long-range impact very pronounced. In fiscal year 1967-1968, local taxes accounted for 44 percent of local revenue, compared to 32 percent for intergovernmental aid. In fiscal year 1976-1977, local taxes accounted for only 38 percent, compared to 39 percent for intergovernmental aid.

It is natural to expect that accompanying this development will be federal and state controls that will inhibit the ability of a local government to develop an independent and long-range personnel policy. The loss of autonomy of New

Table 14-2
Distribution of Local Tax and Intergovernmental Revenues as a Percentage of Total Local Government Revenues, 1967-1977

Year	*Intergovernmental*	*Local Taxes*
1967-68	31.8	44.4
1968-69	32.9	43.9
1969-70	33.1	43.6
1970-71	34.1	43.0
1971-72	34.5	43.2
1972-73	37.1	41.1
1973-74	38.3	39.4
1974-75	38.8	38.4
1975-76	39.1	37.9
1976-77	39.2	38.1

Sources: U.S. Department of Commerce, Bureau of the Census, *Governmental Finance* (1969-1978).

York City to state agencies is a major and self-evident example of this trend. In addition, recent years have seen new controls placed upon programs such as General Revenue Sharing and Community Development Block Grants. These controls constrained even further local government personnel policy options. Finally, this pattern is reinforced even further by the increasing tendency of the federal government to shift intergovernmental revenue from welfare state programs to economic stimulus programs, such as the Comprehensive Employment and Training Act (CETA), which prohibit substitution and long-range finance and support.[5]

Thus, unlike the Johnson War on Proverty programs or even Nixon's New Federalism programs which provided new money with relatively few strings attached to personnel planning, current federal programs substantially limit the options that local management have for making personnel decisions. From the union perspective, the shifts in revenue sources and authority may mean that local collective bargaining is becoming less useful as a mechanism for initiating fundamental changes in wage and working conditions.

In summary, all three of these fiscal pressures are working simultaneously to reduce the ability of administrative decision makers and employee representatives to bargain and to compromise. This combination of pressures confuses decision making, makes long-range planning very difficult, and antiquates previously held personnel policies. The impact and direction of fiscal pressures on the labor-management-relations model has definitely changed in recent years.

Managerial

The managerial environment of local governments has been changing as well. Of crucial importance to the prospects for public-sector employee unionization are recent changes in the power balance between appointed administrative and elected officials. Shifts in this balance are quite important and can affect the entire administrative climate of an organization[6] as well as public employee labor relations. It is my contention that recent shifts in the balance appear to be benefiting elected public officials and to be threatening the power and autonomy of administrators and public-sector unions. In particular, the managerial trends are eroding the monopoly of expertise, the support of client groups, and the crucial services provided by appointed administrators and union personnel.[7] The growth of local-level citizen participation, regional governments, and program management systems are all manifestations of these trends.

Citizen Participation. Citizen participation by itself appears to be a noble and harmless concept. Nevertheless, it has helped augment the power of elected officials vis-à-vis appointed administrators and public-sector unions. Requirements for citizen input are most effectively managed by elected officials over whom the public can exercise direct authority. In theory, citizen participation suggests direct involvement of citizens in the decision-making process, but in practice it has meant a greater involvement and participation of elected officials in the decision-making process with a commensurate decline in the legitimacy of municipal administrators and unions. The delegation of responsibility for general policy from the administrator to the elected official cannot but set a precedent for behavior in other areas. Thus personnel policy and personnel decision, though not under the rubric of citizen participation, are affected by it. Citizen participation, in effect, weakens the client connections and support of the administrator.

From the union perspective, citizen participation and increased populist-type government have eroded union policymaking power and its ability to pressure political figures. The increasing frequency of local bonding and revenue campaigns in which administrators and public employee associations find themselves allied against citizen action councils in a generally losing effort, demonstrates the extent to which the political climate has changed.[8]

Regional Government. Associations of governments, planning commissions, or councils of governments have also upset the balance between appointed administrators and elected officials. They have grown dramatically in size and scope in recent years. For example, the 1977 Census of Governments listed 1,569 regional organizations, nearly one-half of which (43 percent) were founded since 1970.[9] These organizations are generally organized to provide professional assistance to local governments and to provide a forum in which governments can coordinate policy. The significance of their growth to our concerns, however, is that these regional organizations, though providing a vitally needed service, tend to weaken the monopoly of expertise held by locally appointed administrators. In this manner, they contribute to a shifting of the administrator/elected official balance. Finally, intergovernmental regional bodies can contribute to a rigidity and standardization in personnel policy that may also have an adverse impact upon the future of labor relations in local governments.

From the union vantage point, regional governments provide a function analogous to the employee association in private industry. These governments can potentially help lessen the threat from a strike by providing alternate services in the event that a strike occurs. They can and have broken the supposed monopoly of power enjoyed by the union, and they have greatly altered the labor-management-relations model.

Program Management Systems. The growth on the local level of management by objective programs, budgeting schemes like planning programming budgeting system (PPBS) or zero-based budgeting, represents a third trend that has adversely affected the balance of administrators and elected officials. To the extent that management by objectives (MBO) and other schemes are implemented, elected officials can gain control over programs and policy and can weaken administrator control over these areas. If nothing else, the potential increases the possibility for a major reshuffling of personnel and for major alterations in personnel policy. In fact, one of the major advantages such programs offer the elected official is a promise of control over the direction and growth of personnel policy.

In summary, managerial changes, unlike fiscal changes, tend to reinforce each other, and to place potentially severe constraints on the ability of a municipal administrator to bargain for a local government, conduct personnel policy, and control labor relations. It appears that in a period when fiscal controls are becoming increasingly tight, the administrator's own latitude is becoming more constrained as well. The union is also affected. The managerial changes have reduced their own political power, helped realign employees and administration together, and threatened their own monopoly control over services. These adverse changes coupled with severe fiscal constraints may be responsible for major shifts in the direction, scope, and character of labor-market relations in the local sector.

Demographics

The third major area of change in the systemic environment of municipal and county labor relations is demographic. The demographic changes, especially in terms of population movement and population structure, are quite pronounced and can have a very substantive impact upon the climate of local governmental labor relations. Most importantly, however, unlike the other two areas of change, these changes are largely outside the control of local actors.

Population Migration. One of the most important trends in population distribution is the reversal of the growth in metropolitanization. While many large metropolises continue to grow, they are now beginning to grow at a slower rate than small towns and communities. Central cities have fared worst. For example, thirty-four of the fifty largest cities registered population losses from 1970 to 1975.[10] These population trends directly affect the demand for services, causing a decline in demand in some areas with a commensurate overstaffing and decline in productivity of the local government labor force.[11] In many public service areas, personnel size is not elastic below certain levels and attempts to reduce or redistribute personnel are resisted.[12] New York City is a prime example of this trend in operation.

Shifts in Population Structure. A second major shift in the population that has occurred since 1970 may have severe repercussions upon county and municipal labor relations. It has only now begun to be felt. This is the shrinkage in the size of the average family from 3.7 children in 1957 to 1.8 in 1975. The shrinkage is affecting the demand for public services, especially in areas like education where personnel structure is not elastic and cannot readily adapt to changes in demographic characteristics. Demands for housing, parks, recreation, and health services are also affected by these shifts.

The combined effect of these fiscal, managerial, and demographic changes in the systemic environment of local government is pronounced. New fiscal constraints imply a decline of local control over revenues and expenditures. New management constraints suggest less room for maneuvering and compromising. The demographic changes create pressures to reduce or redistribute services and the personnel who provide them. Together, the systemic constraints suggest that serious problems may develop in the local public personnel sector, and especially in public employee unionization.

Implications of Structural Trends on Labor Relations

The structural trends discussed above can have a substantive and profound impact on the future of public employee unionization in local government. These trends can decrease the potential for growth of public employee unions, contribute to the deterioration of the bargaining climate between cities/counties and their employees, and help turn public opinion against public employee collective action.

Impact upon Growth

In recent years, public employee union membership has grown dramatically so that in 1976, 54 percent of all local employees were members of employee associations and nearly 41 percent were associated with collective bargaining units.[13] The trend in growth, however, has slowed in the last several years, and there may be an absolute decline expected as the large cities and northern industrial local government work forces decline in size. To illustrate this point, a larger percentage of local employees were organized in 1974 than in 1976, and twenty-five states registered declines in membership in 1975 compared to 1974 and sixteen states declined in absolute membership from 1975 to 1976.[14] Many of these states were northern or heavily industrialized.

The regional disparity in public employment growth may explain part of the slackening growth rate. Contrary to popular legend, the percentage of the population employed by local governments has grown at a much more rapid rate in the sunbelt states than in the older industrialized ones. In some sunbelt states the absolute percentage of public employment is higher than the national average and higher than in the older industrialized areas. Table 14-3 provides summary data from six representative states from 1969 to 1977 illustrating this point.

The implication from this finding is that right-to-work states, southern states, and nonmetropolitan areas which are negatively related to the growth of public-sector unions (Moore 1977) are the same areas which are experiencing the growth in public sector employment on the local level. Therefore the prospects for future growth of these unions, given the demographic changes in the country, do not seem very bright. Obviously, to survive, public employee unions are going to be forced to engage in protracted organizational drives involving extended battles for recognition.

Second, the trend toward contracting out basic services such as sanitation and even fire protection as a mechanism for cutting costs and personnel will

Table 14-3
Local Public Employment per 10,000 Population for Selected States and National Average, 1969-1977

		Old Industrial			*Sunbelt*		
Year	*National Average*	*Calif.*	*Conn.*	*Mich.*	*Georg.*	*N.C.*	*Texas*
1969	296.1	345.5	241.5	298.5	288.5	243.1	286.6
1971	311.3	355.7	252.0	304.3	323.8	269.2	312.9
1973	335.0	372.0	288.0	315.0	355.0	288.0	343.0
1975	347.0	393.0	289.0	350.0	376.0	296.0	355.0
1977	351.0	391.0	288.0	340.0	389.0	347.0	364.0

Sources: U.S. Department of Commerce, Bureau of the Census, *Public Employment in (1970-1978).*

put the public employees and their associations under severe pressure and contribute even more toward the development of a siege mentality on the part of the membership.[15] In this atmosphere, relatively small issues will escalate and the quality of public-sector labor relations will deteriorate. This change in perception will increase the probability of longer, larger public employee strikes and reawaken the threat of breaking union power in municipal and county government. While reliable statistics do not exist on the extent of contracting out, the discussions in city councils and county boards of commissioners are widespread and serious. If this trend continues, serious labor conflict centering around recognition and seniority is inevitable.

Impact upon the Bargaining Climate

The change in the systemic environment also can be expected to have a pronounced impact upon the climate of public-sector collective bargaining. It is very likely that the incidence of multilateral collective bargaining will increase and that the legalistic structure in which collective bargaining must be conducted will be altered in many areas.

The occurrence of multilateral collective bargaining is strongly affected by the relative power position of administrative and elected officials. Management in the public sector is difficult to identify, and it is even more difficult for the appointed managers and elected officials to achieve a united and mutually trusting outlook and bargaining strategy (Wellington & Winters 1971). We can expect, especially in circumstances where the administrator/elected official power balance has shifted, that the ability of the administrator or professional personnel director to conduct labor negotiations will be hampered by the presence and activity of elected public officials.

In the past, multilateral collective bargaining was perceived as a boom for labor (Kochan 1974, 1975). Today, however, such a situation could weaken considerably labor's bargaining position. The major difference now is that political interference will tend to be antispending, antiexpansionist, and antibureaucratic. Therefore an increase in multilateral collective bargaining, resulting from changes in the managerial environment, will weaken organized labor's bargaining position as it finds its demands and tentative agreements rejected by higher managerial or political levels (Cohen 1979). Of necessity, the result will be more labor unrest, more intense public employee strikes, and poorer public personnel policies.

A second change in the overall bargaining climate is the legal structure of public employee negotiations. It is unlikely that the trend toward greater legislation of a public employee's right to strike will continue. Events such as the attempted 1978 Missouri referendum to repeal the closed shop indicated the depth and direction of the changing bargaining climate. For example, no major municipality, with the possible exception of Atlanta, has recognized the public employee union's right to bargain collectively in recent years. More typical of the resistance to legal change is the hapless and tragic example of Memphis where police and fire personnel struck. In fact, more than a quarter of all public employee strikes in 1976 did not end with a formal settlement; this record indicates that there is considerable resistance to collective bargaining negotiations in the public sector and that future gains will not come easily.[16]

The systemic trends discussed earlier are probable causes for sharpened resistance to collective bargaining. For example, Gryski (1978) noted that the protection of public employee rights is strongly associated with overall union recognition, higher-wage salaries of state and local employees, higher-education levels, higher-family incomes, greater population density, and propensity to innovate. All these associations, however, are inversely related to the fiscal, managerial, and demographic trends discussed earlier. The future therefore suggests considerable resistance to public employee collective bargaining attempts and a continuing lack of legal protection for such attempts. By inference, this projection means that when disputes in public employee labor relations occur, they will tend to occur at a much higher level with little probability of a speedy and equitable settlement.

Impact upon Public Opinion

The third major impact of the systemic changes upon public-sector labor relations is on the climate of public opinion. This climate is tied in and integrally linked with such factors as growth and bargaining rights. The demographic changes, for instance, indicate that public employee unionization drives must organize indigenously hostile areas if they wish to expand. Such attempts will not be popular in the affected areas or in the mind of the general public. The growing fiscal and managerial constraints suggest that public employees, especially those in nonvital services, will be the first to feel the pressure of the public. Finally, the weakening of administrative personnel in favor of the elected officials removes an effective protective buffer between the public employee and the public. The result is an emotional climate in which the public servant is alienated from his community and the community perceives its employees as parasites or public scavengers, a prospect that is not at all encouraging.

Summary

The components of the systems model in which labor-management relations is conducted have shifted markedly since the late 1960s and early 1970s. The old threats of government by union, political blackmail by unions, and union dominance over public policy issues never materialized. Instead, the situation today is in a greater degree of flux and uncertainty. Individual administrators and union officials have much less control over events than was the case in the past. Victory for one side or another is difficult to achieve or even identify. Management, political, and employee roles are less defined as well.

Finally, the potential for labor conflict is heightened since the viability of the public-sector unions is threatened. In addition, the bargaining climate is being muddied and politicized, and the communities seem to be crying for revenge. It is a situation in which there can be no winners. Perhaps this suggests the beginning of the maturation of labor-management relations in the public sector.

Notes

1. David T. Stanley, *Managing Local Government under Union Pressure* (Washington, D.C.: Brookings Institution, 1972): 136. Reprinted with permission.

2. Zagoria (1972) and Stanley (1972) are exceptions. I have also excluded most of the extremist opinions on this subject since they lack scholarly objectivity.

3. Implications of Structural Trends on Labor Relations in this chapter provides evidence that substantiates these claims.

4. U.S. Department of Commerce, *Governmental Finance in 1977* (Washington: U.S. Government Printing Office, 1978).

5. See John J. Kirlin (1978) for an excellent analysis of the implications of this trend for local governments.

6. Riggs (1969) presents a description and analysis of the importance of the balance.

7. The terms and concepts are borrowed from Rourke (1976).

8. Oberer (1969) predicted this very well.

9. U.S. Department of Commerce, *1977 Census of Governments* (Washington: U.S. Government Printing Office, 1978), p. 7.

10. U.S. Department of Commerce, *County and City Data Book, 1977* (Washington: U.S. Government Printing Office, 1978).

11. Ehrenberg (1972) traces the impact of these changes.

12. See Ladd (1977) for an empirical test of this proposition.

13. U.S. Department of Commerce, *Special Studies no. 88* (Washington: U.S. Government Printing Office, 1978), pp. 8-9.

14. Department of Commerce, *Special Studies no. 81, 88* (Washington: U.S. Government Printing Office, 1977).

15. The policy position of the American Federation of State, County, and Municipal Employees (AFSCME) concerning the Proposition 13 referendum issue in California illustrates the viability of this perception on the part of organized labor.

16. U.S. Department of Labor, *Analysis of Work Stoppages, 1976.*

References

Anderson, John C. "Bargaining Outcomes: An IR System Approach." *Industrial Relations* 18, no. 2 (Spring 1979): 127-143.

Bakke, E. Wight. "Reflections on the Future of Bargaining in the Public Sector." *Monthly Labor Review* 93, no. 7, (July 1970): 21-25.

Benecki, Stanley. "Municipal Expenditure Levels and Collective Bargaining." *Industrial Relations,* 17, no. 2 (May 1978): 216-230.

Burton, John F., Krider, Charles E. "The Incidence of Strikes in Public Employment." In Hamermesh, Daniel, ed., *Labor in the Public and Nonprofit Sectors.* Princeton, N.J.: Princeton University Press, 1975, pp. 135-177.

Cohany, Harry P. and Dewey, Lucretia M. "Union Membership among Government Employees." *Monthly Labor Review* 93, no. 7 (July 1970): 15-20.

Cohen, Sanford. "Does Public Employee Unionism Diminish Democracy?" *Industrial and Labor Relations Review,* 32, no. 2 (January 1979): 189-195.

Couturier, Jean J. "Crisis, Conflict and Change: The Future of Collective Bargaining in Public Service." In Murphy, Richard J. and Sackman, Morris, eds., *The Crisis in Public Employee Relations in the Decade of the Seventies.* Washington, D.C.: Bureau of National Affairs, 1970. pp. 115-121.

Ehrenberg, Ronald G. "Municipal Government Structure, Unionization, and the Wages of Fire Fighters." *Industrial and Labor Relations Review* 27, no. 1 (October 1973): 36-48.

———. *The Demand for State and Local Government Employees.* Lexington, Mass.: Lexington Books, 1972.

Gryski, Gerard. "Factors Related to Public Employee Union Strength." Paper presented at the annual meeting of the American Political Science Association, New York, September 1978.

Hale, George E., Palley, Marian Lief. "Federal Grants to the States." *Administration of Society* II, no. 1 (May 1979): 3-26.

Kirlin, John J. "Structuring the Intergovernmental System: An Appraisal of Conceptual Models and Public Policies." Paper presented at the annual meeting of the American Political Science Association, New York, September 1978.

Kochan, Thomas. "City Government Bargaining: A Path Analysis." *Industrial Relations* 14, no. 1 (February 1975): 80-101.

———. "A Theory of Multilateral Collective Bargaining in City Governments." *Industrial and Labor Relations Review* 27 (July 1974): 525-542.

Ladd, Helen. "An Economic Evaluation of State Limitations on Local Taxing and Spending Powers." Department of City and Regional Planning, Harvard University, Cambridge, Mass., 1977.

Love, Thomas and Sulzner, George. "Political Implications of Public Employee Bargaining." *Industrial Relations* 11 (February 1972): 18-33.

McCarthy, Charles F. "Collective Bargaining and the Local Chief Executive." *Public Personnel Review* 31 (July 1970): 157-161.

Moore, William J. "Factors Affecting Growth in Public and Private Sector Unions." *Journal of Collective Negotiations in the Public Sector* 6, no. 1 (1977): 37-43.

Oberer, Walter E. "The Future of Collective Bargaining in Public Employment." *Labor Law Journal* vol. 20 (December 1969): 777-786.

"Public Employees Lose Leverage, Job Security and Job Rights are Threatened by the Spending Pinch," *Business Week,* December 22, 1975, p. 15.

Riggs, Fred. "Bureaucratic Politics in Comparative Perspective." *Journal of Comparative Administration* 1, no. 1 (1969): 5-39.

Rourke, Francis E. *Bureaucracy, Politics and Public Policy* 2d. ed. Boston: Little, Brown, 1976.

Slatzstein, Alan L. "Can Urban Management Control the Organized Employee?" *Public Personnel Management* 3 (July-August 1974): 332-340.

Stanley, David T. *Managing Local Government Under Union Pressure.* Washington: Brookings Institution, 1972.

Toledano, Ralph de. *Let Our Cities Burn.* New Rochelle, New York: Arlington House, 1975.

U.S. Department of Commerce, Bureau of the Census. *County and City Data Book,* 1977. Washington, D.C.: U.S. Government Printing Office, 1978.

———. *Governmental Finance* GF Series no. 5. Washington, D.C.: U.S. Government Printing Office, 1969-1978.

———. *Labor Management Relations in State and Local Governments: 1976.* Special Studies no. 88. Washington, D.C.: U.S. Government Printing Office, 1978.

———. *Labor Management Relations in State and Local Governments: 1975.* Special Studies no. 81. Washington, D.C.: U.S. Government Printing Office, 1977.

———. *Public Employment* in GE Series, no. 1. Washington, D.C.: U.S. Government Printing Office, 1970-1978.

———. *Statistical Abstract of the United States 1977.* Washington, D.C.: U.S. Government Printing Office, 1977.

———. *Taxes and Intergovernmental Revenues of Counties, Municipalities and Townships: 1976-1977.* GF 77, no. 9. Washington, D.C.: U.S. Government Printing Office, 1978.

———. *Taxes and Intergovernmental Revenues of Counties, Municipalities and Townships: 1975-76.* GF 76, no. 9. Washington, D.C.: U.S. Government Printing Office, 1977.

———. *1977 Census of Governments, Regional Organizations.* vol. 6. Washington, D.C.: U.S. Government Printing Office, 1978.

U.S. Department of Labor, Bureau of Labor Statistics. *Analysis of Work Stoppages, 1976.* Bulletin 1996. Washington, D.C.: U.S. Government Printing Office, 1978.

Wellington, Harry H., and Winter, Ralph K. *The Union and the Cities.* Washington, D.C.: Brookings Institution, 1971.

Wright, Deil S. *Understanding Intergovernmental Relations: Public Policy and Participants' Perspectives in Local, State and National Governments.* North Scituate, Mass.: Duxbury, 1978.

Zagoria, Sam. "The Future of Collective Bargaining in Government." In Zagoria, Sam, ed., *Public Workers and Public Unions.* Englewood Cliffs, N.J.: Prentice-Hall 1972. pp. 160-177.

15 Public-Sector Unionism and Tax Burdens: Are They Related?

Jeffrey D. Straussman and
Robert Rodgers

One factor that has affected state and local politics since the early 1960s is the growth of public-sector unionism. This growth has occurred along three dimensions: the militancy of public workers as reflected in strike activity, legislative recognition of public unions as a political force in the form of collective bargaining laws, and the increase in earnings of public employees. Specialists of public-sector labor relations have labeled the 1960s as the "first generation" of public-sector unionism where the political power of unions focused on catch-up wage gains.[1]

The expansion of the 1960s gave way to the more restrictive 1970s as dramatized by the near financial collapse of the City of New York in 1975 and the "taxpayers' revolt" of 1978 symbolized by California's Proposition 13. In searching for explanations and scapegoats for the fiscal ills of state and local governments, many politicians and political observers have assumed that the first-generation catch-up phase of public unions of the past decade has contributed heavily to the current fiscal deterioration afflicting many states and municipalities. For example, in 1976 Governor Carey of New York wanted to suspend bargaining with public unions as a way to convince investors that the state was willing to resist wage increases as part of an effort to resolve its fiscal problems.[2] More generally, public-sector labor-relations specialists David Lewin, Peter Feuille, and Thomas Kochan point out that, "the Democratic governors of the states of California, Wisconsin, Illinois, New York, Connecticut, and Massachusetts . . . all have adopted positions which effectively reduce the political access of influence of public unions that helped put them in office."[3]

While the assumption that the growth of public-sector unionism (as indicated by strikes, legislation, and earnings) has exacerbated fiscal deterioration certainly reflects conventional wisdom, the relationship has not been examined empirically. The purpose here is to provide a preliminary investigation of the relationship of public-sector unionism on interstate variations of fiscal conditions across the fifty states using 1960-1972 as a period of analysis. It is our view that public unions may be receiving more blame for the fiscal condition of many municipal and state governments than they perhaps deserve. To demonstrate this, we focus on one dimension of the fiscal climate of states: the tax burden.

Richard Block, Robert Jackman, and Brian Silver provided helpful comments on earlier versions of this work.

Our general model of the determinants of the tax burden (TB_{it}) of state and local governments assumes:

$$TB_{it} = F(\overline{UN}_{it}, \overline{NU}_{it})$$

where TB_{it} is a measure of the tax burden in jurisdiction i at time period t. $\overline{UN}_{it}$ denotes a vector set of variables measuring activity levels of public-sector unionism in jurisdiction i at time period t. $\overline{NU}_{it}$ denotes a vector set of nonunion determinants of the tax burden in jurisdiction i at time period t. While we are primarily interested in the impact of public-sector unionism on differences in the tax burden across the states, the consideration of nonunion factors is important since even the harshest critics of public unions would recognize that the fiscal climate of a state might be attributed to conditions that have little or nothing to do with public-sector unions. Consequently, the selected variables in vector set $\overline{NU}_{it}$ should have theoretically independent or combined impacts on the tax burden.

Tax Burden: The Dependent Variable

While there is no simple indicator of the financial health of state and local governments, one dimension of a state or local government's fiscal condition is the tax burden as measured by revenue efforts.[5] If state spending increases as a result of new demands for goods and services or perhaps because public employees bargained successfully for wage gains, revenue must increase accordingly. However, given the fact that elected officials are generally loath to raise taxes, particularly if it is seen that an increase in taxes will interfere with a candidate's electoral chances,[6] one might expect a lag between new pressure for state spending and additional revenue to pay for the increase. We tapped the revenue response of state governments as an annual mean change in tax effort (ΔTE) from 1960 to 1971.[7]

Public-Sector Unionism

Public-sector unionism (the $\overline{UN}$ vector set) is identified with three indicators: the policy taken toward public unions as reflected by the comprehensiveness of state-level public-sector collective bargaining legislation (L), the militancy of public unions as demonstrated by strike activity (S), and the earnings (E) of public employees.

The relationship between earnings and the tax burden is the most straightforward. Given that labor costs comprise approximately 70 percent of a municipal budget and about 40 percent of a state budget, an increase in public-sector wages will increase government spending. In addition, government has relatively little flexibility to affect the cost of supply of goods and services by substituting capital for labor; therefore, in the short run wage increases put direct pressure on the public budget and, as a result, are expected to be positively related to a change in tax effort.[8] While this hypothesis has not been examined directly, Thomas Muller has shown that declining urban areas have higher labor costs than growing urban areas.[9] The relationship between wages and the change in tax effort was examined by using mean monthly earnings for state and local public employees from 1965 to 1970.[10]

The relationship between the other two indicators of public-sector unionism and the tax burden is less straightforward. Although public unions focused on catch-up wage gains during the 1960s, it is plausible to argue that strike activity and collective bargaining legislation should also have direct effects on the fiscal condition of states. Specifically, unions also strike and bargain for nonwage outcomes such as sick leave, pensions, insurance, programs, work breaks, and the like.[11] Nonwage outcomes put pressure on the public budget and, like wages, require offsetting revenues if fiscal decline is to be avoided. To examine the effects of both militancy and collective bargaining legislation, we adopted a measure of the number of strikes per ten state and local government employees, and constructed an index of the comprehensiveness of state-level collective bargaining legislation.[12] Each is predicted to be positively related to the change in tax effort.[13]

Nonunion Factors

We selected nonunion factors ($\overline{NU}$ vector set) for this analysis based on the following proposition: the enlargement of governmental activity should tend to worsen the tax burden. As noted previously, this assumption is based on the expectation that elected officials are generally reluctant to increase taxes. While governments of course do increase taxes, we might, at the least, expect some lag between increased governmental activity which translates into a growth in public spending, and a corresponding growth in revenue from own sources. Given this formulation the next task is to select indicators of increased government activity.

To accomplish this objective we first borrowed two well-known concepts from the voluminous comparative state politics literature: political competitiveness and innovation. Although both concepts have sparked controversy in the field of comparative state politics, particularly their ability to explain variations in public policy, they are used in the present analysis for a different purpose. Specifically, it is hypothesized that both political competitiveness and innovation are correlates of expanded government activity and therefore should be positively associated with an increase in tax burdens. We use the Ranney index (R)[14] and the Walker innovation index (W) to test this association here.

Second, we further assume that one impact of interest-group pressure is the passage of legislation that furthers the immediate interests of group members or broader political concerns. Such legislation should produce upward pressure on expenditures thereby exacerbating any expenditure-revenue gap. While this hypothesis should apply to all groups that have legislative demands, we singled out organized labor in the private sector. Organized labor in the private sector has generally supported expanded programs that directly affect union members and their families. In addition, organized labor in the private sector has also been supportive of public-sector unions. We expect a positive relationship between the percentage unionized in the private sector (I) and the change in tax effort.

In this analysis we focused on two environmental conditions that should affect the tax burden: the change in population from 1960 to 1972 and the average unemployment rate from 1965 to 1970. Although there is some debate on the issue, we hypothesize that a change in population should be negatively associated with the tax burden. One reason for this expectation is that economies of scale in the provision of goods and services mitigate the potential rise in per capita spending. Moreover, any increase in spending as a result of population growth should be offset by additional revenues from the same population.[15] Conversely, states that have lost population since 1960 are faced with an increasingly dependent population and a shrinking tax base. In other words, the change in population indicator captures the regional patterns expressed in the phrase sunbelt growth and frostbelt decline. This decline is presumably also indicated by fiscal decline, or so the conventional wisdom goes. We also assume that unemployment puts pressure on the budget because at least a portion of the unemployed require government services that they would not receive if they were employed. Public assistance is a good example. In addition, unemployment obviously reduces the taxes collected and should increase the tax burden for citizens in states with high unemployment. We therefore expect the mean unemployment rate to have a positive impact on the change in tax effort.

Analysis

To examine interstate variations in tax burden we first enter the variables selected to test the model in a multiple regression equation. The results are as follows:

$$\begin{aligned} \Delta TE = &- .0031\,L - .0275\,S + .0132\,E - .3814\,U \\ &\quad (0.16) \qquad (0.20) \qquad (3.53) \qquad (2.45) \\ &- .0478\,P - 7.175\,W + 2.939\,R - .0254\,I \\ &\quad (4.74) \qquad (2.59) \qquad (2.08) \qquad (1.31) \end{aligned} \tag{15.1}$$

Earnings (E), percentage of change in population (P), unemployment (U), Ranney index (R), and Walker innovation index (W) have significant T-scores (indicated in parentheses) at the .05 level; however, the coefficients on U and W do not have the expected signs. Most important, equation 15.1 demonstrates that only one of the three indicators of public-sector unionism (earnings) is significantly different from zero. The explanatory power of the equation is nonetheless good, with an adjusted $\bar{R}^2$ of .44 and an F-score of 5.56.

While two of the three public-sector union factors selected for examination in this study do not have independent effects on the change in tax effort, these same measures might exhibit *combined* effects. For example, in another paper we treated percentage of change in population, mean unemployment rate, and a third environmental variable—average percentage of urbanization—as indicators of demand for governmental goods and services. In view of the relatively low unemployment and the tight labor market during the time period under investigation, we reasoned that public employees were in a good position to obtain wage gains from state and local governments. Yet while all three indicators of demand were positive and significant and also explained much of the variation in the earnings of public employees, the indicators of demand in interaction with earnings failed to contribute to any understanding of interstate variations in fiscal deterioration as measured by a change in tax effort.[16]

Since we are primarily interested in exploring the conventional wisdom that public unions have increased the tax burden, interaction effects of public-sector union indicators with nonunion factors remain an important consideration. For example, if private unions exert pressure on state legislators to support the interests of public employees, this should be reflected in more comprehensive state-level public-sector collective bargaining legislation. Therefore it seems appropriate to specify an interaction ($L \times I = LI$) between the percentage unionized in the private sector (the interest-group variable) and the comprehensiveness of state-level public-sector collective bargaining legislation. This line of reasoning can be extended to include a joint impact (LR) of the comprehensiveness of state-level public-sector collective bargaining legislation and the Ranney index (the most competitive states are likely to have higher scores on the legal index), and a joint impact (LW) of laws and the Walker innovation index (innovative states are more likely to pass more comprehensive public-sector collective bargaining legislation).

This alternative specification states that, with the exception of earnings, there are no direct relationships between the public unionism and the change in tax effort. Rather, variation in ΔTE might be better explained by these combined effects. To test this possibility a second regression included the interactions specified above (LI, LR, LW). The results are as follows:[17]

$$\begin{aligned}\Delta TE = \; & \underset{(0.20)}{.0041\,L} - \underset{(0.16)}{.0231\,S} + \underset{(3.31)}{0.129\,E} - \underset{(2.26)}{.3858\,U} \\ & - \underset{(3.63)}{.0426\,P} - \underset{(2.29)}{6.856\,W} + \underset{(0.911)}{1.886\,R} - \underset{(1.22)}{.0246\,I} \\ & + \underset{(0.48)}{.1447\,LW} - \underset{(0.40)}{.1064\,LR} - \underset{(0.70)}{.0019\,LI}\end{aligned} \tag{15.2}$$

The coefficient on earnings is still positive and significant as was the case with equation 15.1, but we find no evidence that the comprehensiveness of state-level public-sector legislation has either a direct or combined impact on the change in tax effort. None of the coefficients on the interaction terms are significant and only one is in the direction predicted. While the explanatory power of this equation ($\overline{R}^2$ = .41) is similar to the adjusted $\overline{R}^2$ attained in equation 15.1, the F-score is reduced to 3.94.

Is public-sector unionism simply unrelated to interstate variations in the tax burden? Clearly the earnings of public employees *are* associated with the fiscal environment of state governments. Notwithstanding the fairly strong relationship here it is still premature to dismiss the conventional wisdom. It is possible that both the statistically and theoretically insignificant union and nonunion determinants indirectly affect interstate variations in the tax burden. Specifically, they may act *through* earnings rather than have an independent effect. A causal model to test this possibility is presented in figure 15-1.

We tested this model using path analysis. The path coefficients indicated in figure 15-2 are the beta weights of the regression coefficients derived from a simple or multiple regression on each variable's causal antecedents. In instances where a variable has only one antecedent, the path coefficient is the simple correlation between the two variables. Notice that in figure 15-2 we have eliminated one insignificant path: the hypothesized relationship between unemployment and earnings.[18]

In general, a comparison of the significance of the path coefficients suggests that the strongest direct and the strongest indirect relationships with ΔTE exist between two of the variables in the nonunion vector set, P and W. In contrast, the path between the comprehensiveness of state-level public-sector collective bargaining legislation and earnings is weak. This finding is also important given the expectation that a zero-order correlation between L and E would be strong and, in this instance, it is (R = .50).[19] Earnings continue to have a direct and significant impact on the tax burden across the states, but earnings in turn are affected by factors which are unrelated to other public-sector unionism characteristics.

Conclusion

Conventional wisdom assumes that public-sector unionism has greatly contributed to the worsening fiscal conditions of many state and local governments. The results of our analysis question this assumption. In particular, we show that the militancy of unions as expressed in strike activity and the policy environment conducive to unionism as reflected in the comprehensiveness of state-level public-sector collective bargaining legislation do not contribute to an explanation of differences in the fiscal condition of states as measured by the change in tax effort from 1960 to 1971.

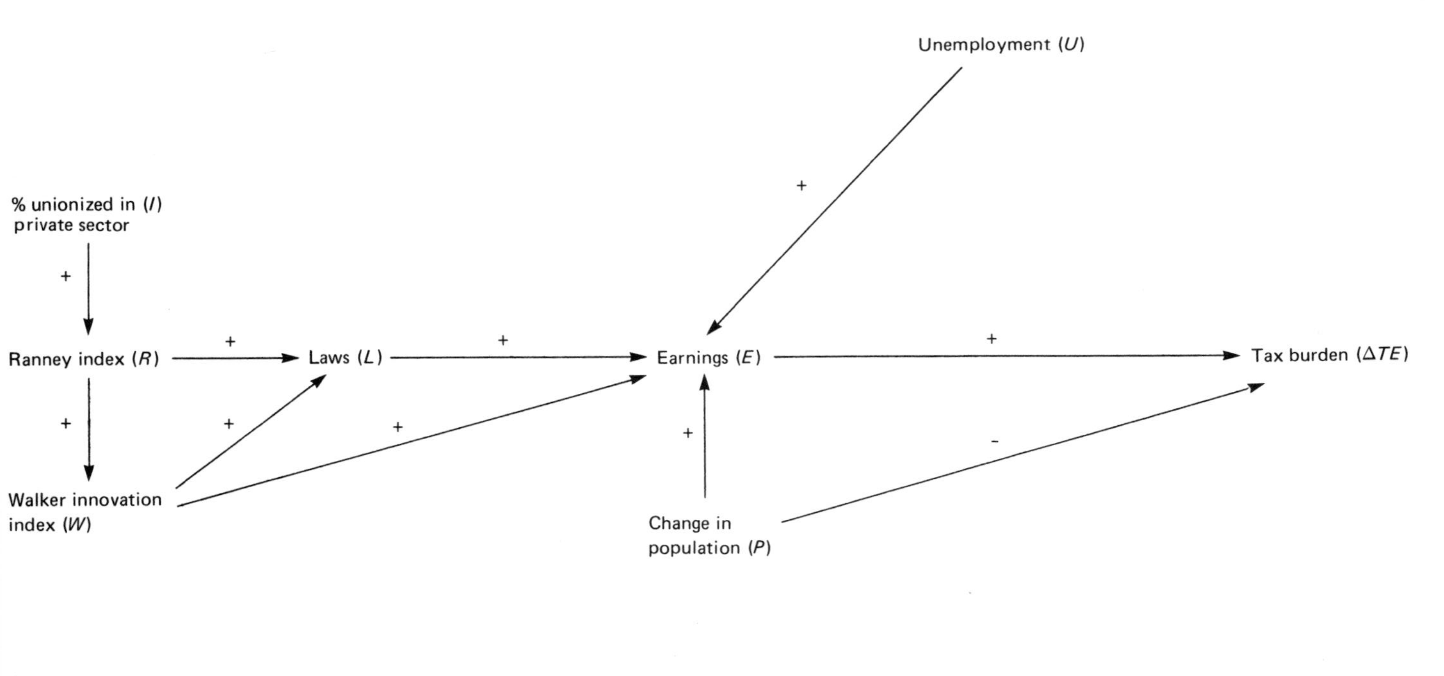

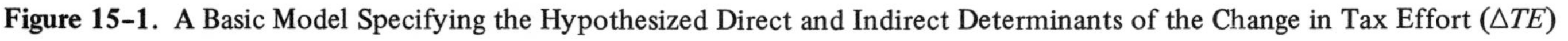
Figure 15–1. A Basic Model Specifying the Hypothesized Direct and Indirect Determinants of the Change in Tax Effort (ΔTE)

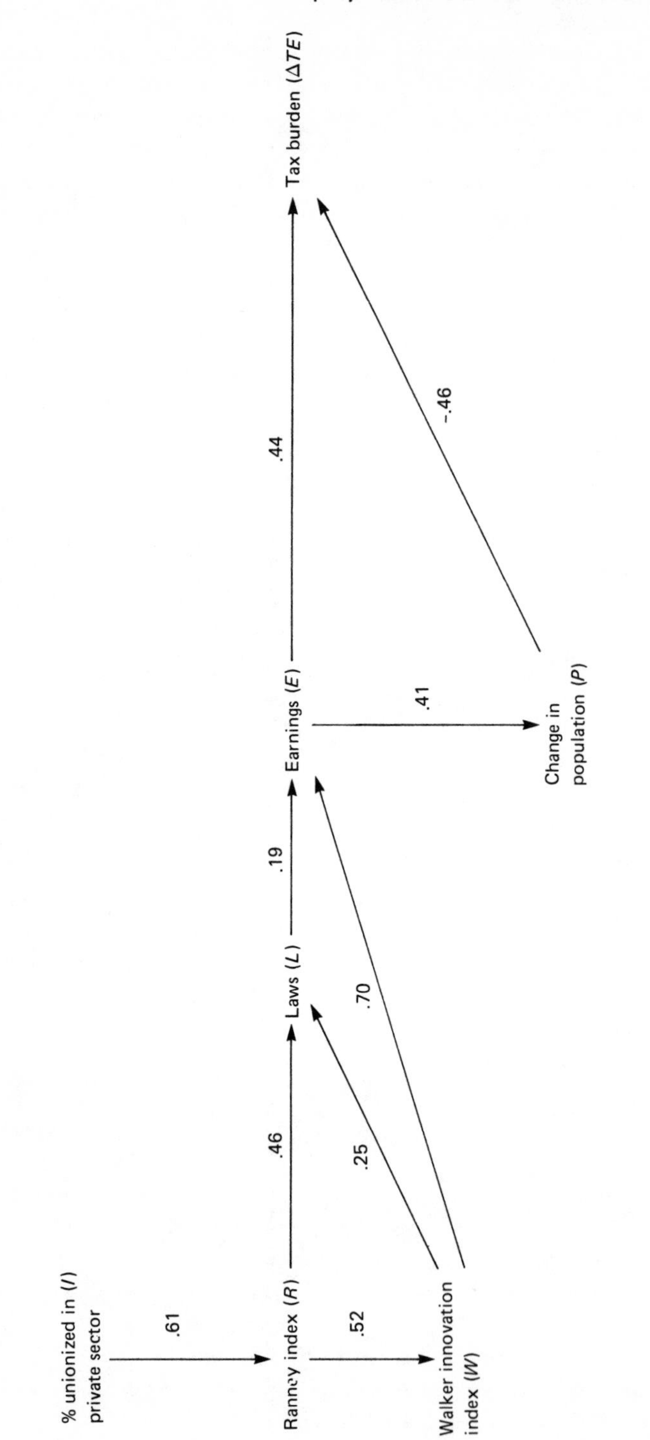

Figure 15-2. Path Coefficients Derived from the Basic Model

Since militancy and policy as determinants of the tax burden faired poorly, public policy debates on public-sector unionism should not be contaminated by these specific dimensions. For example, the passage of more comprehensive collective bargaining laws should not be tied to the (unfounded) expectation that such legislation would worsen the fiscal climate. Of course elected officials are rightly concerned about the impact of wage concessions on tax burdens as our analysis would indicate. But the broader implication of this study is that the earnings of public employees can be attributed to factors that are unrelated to public unionism.

Notes

1. David Lewin, Peter Feuille, and Thomas A. Kochan, *Public Sector Relations* (Glen Ridge: Thomas Horton and Daughters, 1977), p. 361.
2. Ibid., p. 12.
3. Ibid., p. 6.
4. Limited evidence that unions exacerbated fiscal conditions at the municipal level is found in David Stanley, *Managing Local Government under Union Pressure* (Washington: Brookings Institution, 1972), and Terry N. Clark, "How Many New Yorks?" (unpublished ms., 1976).
5. James A. Maxwell and J. Richard Aronson, *Financing State and Local Governments,* 3d ed. (Washington, Brookings Institution, 1977), pp. 38-40.
6. See, for example, Donald H. Haider, "Fiscal Scarcity: A New Urban Perspective," in *The New Urban Politics,* eds., Louis H. Masotti and Robert L. Lineberry (Cambridge, Mass.: Ballinger Publishing Company, 1976), pp. 171-216.
7. The change in tax effort (ΔTE) from 1960 to 1971 is measured as follows:

$$\Delta TE_i = \frac{100}{11} \times \frac{TE_{it_2}}{TE_{it_1}}$$

where ΔTE_i is the average annual percentage of change in tax effort in state i from 1960 to 1971; TE_{it_1} is the tax effort in state i during time period t_1; TE_{it_2} is the tax effort in state i during time period t_2, and t_1 = 1960, t_2 = 1971. Data are from *Handbook of State Policy Indicators, 1960 and 1973.* Unfortunately, this measure of tax effort as constructed by the National Tax Institute is available only for the two points in time, 1960 and 1971.

8. See William J. Baumol, "Macroeconomics of Unbalanced Growth: The Anatomy of Urban Crisis," *American Economic Review* 57 (June 1967): 415-426.
9. Thomas Muller, *Growing and Declining Areas: A Fiscal Comparison* (Washington, D.C.: Urban Institute, 1975).
10. Earnings data are from Bureau of the Census, *Public Employment,* series 1961-1970, table 7.

11. For evidence based on fire fighters see Thomas K. Kochan and Hoyt N. Wheeler, "Municipal Collective Bargaining: A Model and Analysis of Bargaining Outcomes," *Industrial and Labor Relations Review* 29 (October 1975): 46-66.

12. A measure of strike activity (*S*) per 10 state and local government employees was constructed as follows:

$$S_{it} = \frac{10 \times DI_{it}}{EM_{it}}$$

where S_{it} is a measure of public-sector strike activity in state *i* during time period *t*; DI_{it} is the average annual number of days idle of state and local government employees in state *i* during time period *t*; EM_{it} is the average annual state and local employment in state *i* during time period *t*; and *t* is the six-year period from 1965 to 1970.

The comprehensiveness of state-level public-sector legislation covering local employees (*L*) in all fifty states was coded using the earliest comprehensive summary of laws available from the Department of Labor. This index has been adapted from earlier work on state and local collective bargaining laws by Thomas Kochan, "Correlates of State Public Employee Bargaining Laws," *Industrial Relations* 12 (October 1973). *L* is an index that assumes that the more formalized the provisions in the enabling legislation favoring public-sector unionism, the more comprehensive is the legislation. For twelve categories a number was assigned to reflect coverage in the law for that particular provision. The total index then consists of summing up across all twelve categories, so that higher scores are awarded to states with the more comprehensive legislation. For example, the score for New York is 19, for Georgia, 0, and for North Dakota, 9. The maximum value is 24, minimum value is 0; the mean is 9.98 and the standard deviation is 9.02.

Data for strike activity come from U.S. Department of Labor, *Current Status of Public Sector Labor Relations*, 1971, and U.S. Department of Commerce, *Public Employment*, series 1961–1970, table 6. Data for the comprehensiveness of public-sector legislation are from U.S. Department of Labor, *Summary of State Policy Regulations for Public Sector Labor Relations: Statutes, Attorney General's Opinions and Selected Court Decisions*, 1971.

13. This prediction is further supported by a finding of a positive and significant relationship between the comprehensiveness of a state public employee bargaining law and bargaining outcomes in municipal governments. See Thomas A. Kochan and Hoyt N. Wheeler, "Municipal Collective Bargaining: A Model and Analysis of Bargaining Outcomes," *Industrial and Labor Relations Review* 29 (October 1975): 46-66.

14. The Ranney index was inverted so that the maximum score is for the most competitive state.

15. See Thomas R. Dye and John A. Garcia, "The Political Economy of Growth Policies in Cities," *Policy Studies Journal* 6 (Winter 1977): 175-185. For a different view see William J. Baumol, "Urban Services: Interactions of Public and Private Decisions," in *Needs, Services and Financing: Readings on Municipal Expenditures*, ed., Patrick Beaton (New Brunswick, N.J.: Center for Urban Policy Research, 1974), pp. 11-30.

16. Robert Rodgers and Jeffrey D. Straussman, "The Impact of Public Sector Unionism on the Fiscal Conditions of the States," Working Paper no. 2, Public Policy and Policy Studies Center, Michigan State University (June 1978).

17. Multicollinearity among the interactive term predictors was minimized by subtracting the means of L, W, R, and I from their observed values and constructing $L \times W$, $L \times R$, and $L \times I$ accordingly. We tested all possible combinations of the interaction terms. Across all the equations examined there are no meaningful differences in the size and direction of the coefficients from those reported in equation 15.2.

18. The insignificant relationship between unemployment and earnings may be mediated by a third consideration, that is, the differences in federal contributions among the states. Preliminary analysis supports this contention.

19. While an examination of the observed minus the predicted correlations among all the variables in the model suggests that there are nontrivial factors which have been omitted, predictions of the correlations between the tax burden and the other variables in the model provide a respectable fit.

16 Public-Sector Unionization and Municipal Wages: The Case of Fire Fighters

William Lyons and
Russell L. Smith

Introduction

During the last twenty years the growth of public employee unions has been dramatic. The policy implications that the proliferation of these unions imply are varied and complicated. This is particularly true of governmental expenditures, where spending on personnel comprises a greater proportion of total service/activity costs than in the private sector. To the extent that public employee unions are successful in obtaining wage increases for their members, governmental service costs may exceed the usual incremental increases that we expect.

The arguments for and against public unionism have pervaded scholarly and public debate about organization in the public sector for many years.[1] One of the principal arguments against public unions has been that they will increase the costs of government beyond what citizens are willing to pay. Indeed, it has been suggested that exorbitant union wage increases have already contributed greatly to the financial bankruptcy of some municipalities, with New York being the most notable example.[2]

Fortunately, a number of researchers have begun to address this question by examining the impact of unionization upon public-sector wages and expenditures.[3] At single points in time we now know that the wage differential for union and nonunion government employees is not great, with the average union impact being around 5 percent. This is not only smaller than public rhetoric leads us to believe, but the likelihood that public employee unions have contributed to the greatly escalated costs of government in recent years is thus placed in doubt.[4]

Because of the cross-sectional nature of the extant research on union wage impacts, however, we cannot say in a definitive fashion whether public-sector union wage demands and associated activities have contributed to the spiral of governmental expenditures. The lack of longitudinal evidence makes it impossible to determine whether wages in unionized governmental units have increased at a rate greater than in nonunionized governments. It may be that the unionized governments paid higher wages to begin with.

To provide information for this debate, this chapter examines the impact of unionization upon the wages and working hours of municipal fire fighters from 1960 to 1970. The units of analysis are all cities with a population of 25,000 or more in 1960.[5] In addition to assessing the relationship between changes in fire-fighter unionization and changes in the dependent variables, changes in indicators of union strength (contracts and dues checkoff) are also examined. Indicators of city government form and supply and demand factors are utilized to determine the independent and multiple contributions of various factors in explaining wage increases.

Research Perspectives

Public Employee Unionism and Fire Fighters

Although it has only been in recent years that public employee unions have increased both their membership and activities, organizations of public employees have existed for some time. The earliest organizations represented skilled or craft workers, with organizations of sanitation workers, teachers, and fire fighters following this pattern. Early union activities most often centered around the issue of employee control by political machines. Thus the municipal reform movement of the late-nineteenth and early-twentieth centuries had an ally in the public-sector labor movement.

While the first public employee organizations were usually limited to single cities and were unaffiliated with national unions, it was not long until local memberships joined national unions. Fire fighters organized very early, most likely as a result of their long working hours. Unlike most public employees who worked shorter hours than private-sector workers, fire fighters were on duty for an average of twenty-one hours each day, and had only one day off each week. The first fire-fighter affiliate of the AFL was chartered in 1903 in Pittsburgh. By 1918 around eighty AFL federal charters had been given to local fire-fighter organizations.[6]

In 1918 the International Association of Fire Fighters (IAFF) was established, with stated goals of wage increases, a two-platoon fire system, and the elimination of political assessments. Until the 1919 Boston police strike, IAFF growth was rapid, with 180 chartered locals.[7] During the 1920s, however, growth was slow, with the union devoting much of its attention to combating antiunion forces characteristic of the period. Today the IAFF has grown to more than 160,000 members with 1,700 locals.[8]

Fire-fighter unionization presents a good opportunity for exploring the broader issue of the impact of public employee unions upon municipal wages, and hence total expenditures. This is largely because the IAFF might be considered a prototypal public employee union. Several factors can be cited to support this. First, not only is the IAFF a traditional craft union and the oldest all-public employee union, but with well over 160,000 members it ranks as the third largest public union.[9] Second, the IAFF is the only public employee union that has not had to compete with other national organizations. To date, no other unions have demonstrated any interest in organizing fire fighters.[10] This makes it possible for the union locals to focus attention upon their membership rather than devoting time and resources to fighting rival unions and associations. Because of the lack of interorganizational conflicts over representation and the nature of the fire-fighter's work life and the cohesiveness it produces, IAFF locals usually enroll an overwhelming majority of the fire fighters.[11]

Third, as Ashenfelter noted, the demand for fire-fighter services in any given city is probably less elastic than the demand for skilled private-sector labor.[12] This may mean that fire fighters are able to negotiate wage increases with fewer adverse effects on their employment than is true for comparably skilled workers in other areas.

Perhaps the most important reason for studying the impact of fire-fighter unionization upon wages is that fire fighters have traditionally been a politically active group. From the inception of the IAFF to 1968, strikes were not allowed by the union's constitution. As Spero noted, a conciliatory strategy emphasizing political support of sympathetic municipal officials and the development of community support was utilized in pressing for union local demands.[13] Although the no-strike provision has been lifted by the IAFF, the conciliatory or political activism strategy of the fire fighters is still prevalent.[14] Among the additional reasons that can be cited for the fire-fighters' success in political pressure activities are their favorable image projected to the public, the large amount of time their job allows for a variety of activities and discussions, and their general willingness to turn out for electioneering or canvassing.[15] In addition, IAFF locals are very active in AFL-CIO Committee on Political Action activities.

Union Wage Impacts

A number of assessments of public-sector union wage impacts have been made. Most relevant to this project are the analyses of fire-fighter unionization.[16] In this section, we first make several comments that apply to most of the union wage impact studies completed to date, and then we turn our attention to the findings of the fire-fighter unionization studies.

Most assessments of public union wage impacts have utilized cross-sectional research designs. Thus using either states or cities as the unit of analysis, researchers have attempted to assess the wage differential between unionized and nonunionized units of government. In addition to looking at the covariance between unionization (variously defined) and wages, indicators of governmental unit socioeconomic and political characteristics, labor demand-and-supply factors, and other political or unionization factors have been examined. Usually the independent as well as multiple impact of these factors have been assessed through multiple regression analysis. These analyses have been conducted for teachers, police, craftsmen, bus drivers, fire fighters, and all municipal employees. While the indicators used in the studies have often differed, the general conclusion offered has been that union/nonunion wage differentials are not great, and that the union impact has been overstaged in public debate.

This conclusion may not be accurate, however, since the research evidence has been built, without exception, upon cross-sectional data. Inferences drawn from cross-sectional data may be spurious. This is especially critical when the focus of research is upon impacts, which imply some sort of change. The differences or lack of differences between unionized and nonunionized employees' wage rates across governmental units may well have existed prior to unionization. Thus historical factors peculiar to cities which actually unionized may be of import. Only a longitudinal design, in which cities are compared with themselves, can correct for these problems.

Examinations of fire-fighter unionization, like most union wage studies in general, have also been cross-sectional in design. The first empirical assessment of fire-fighter unionization was Ashenfelter's 1971 study.[17] Positing that unionized fire fighters would have higher average hourly wages and annual salaries, and fewer average working hours than nonunionized fire fighters, Ashenfelter found that the impact of unionization during any given year from 1961 to 1966 was not very large. The greatest differences were observed for 1966, with unionization increasing average hourly wages from 6 to 16 percent, reducing average annual duty hours by 3 to 9 percent, and increasing the average salary from 0 to 10 percent.[18]

Ehrenberg, in a 1973 study, noted weaknesses in Ashenfelter's analysis and extended it in several ways.[19] Let us look at each of the improvements made by Ehrenberg. First, in addition to utilizing a measure of whether or not a municipality had an IAFF local, a measure of whether a formal contract covering fire fighters existed was also used. Thus Ehrenberg was able to tap what he thought was a union's strength. Second, rather than merely examining average salaries, entry-level and maximum salaries for each city were used as dependent variables. Unions attempt to affect salary scales rather than the "average" values for fire fighters. Third, rather than assuming that the demand for fire-fighter services is perfectly inelastic, Ehrenberg attempted to specify and control for certain fire-fighter demand-and-supply functions. Finally, he attempted to ascertain the impact of city government structure upon the union-wage relationship; a factor ignored by Ashenfelter and of no small theoretical import.

Like Ashenfelter, Ehrenberg found that the existence of an IAFF local did not greatly affect fire-fighter wages and hours of work. Instead, demand-and-supply indicators were most important. The union contract measure produced different results, though supply-and-demand factors were still most important. For cities with union contracts, wages were found to be from 0 to 9 percent higher than in noncontract cities, and hours were found to be 2 to 9 percent less. Form of city government had little or no impact upon fire-fighter wages.[20]

While the Ehrenberg analysis improved upon Ashenfelter's work, the problems accompanying cross-sectional designs still remain. We do not know the extent to which unionization affects fire-fighter wages or working hours. To realistically model the process by which this happens requires that we observe changes in unionization, wages, hours, and other factors for the same city over time. The task of this chapter is thus to more accurately assess the impact of fire-fighter unionization. This is done through a longitudinal design.

Research Design

Modeling the Effects of Unionization

A number of factors may influence fire-fighter wages and working hours. To make unbiased estimates of the effect of unionization across cities, historical influences which may have influenced the hours worked and entry-level salaries of fire fighters must be isolated and removed. These prior effects can be removed by concentrating on changes which occur in a city apart from those which might be expected from past trends for each dependent variable. Variables which might have "caused" earlier values of the dependent variable to be high do not, then, have to be controlled for.

To realistically assess the impact of unionization upon wages and working hours, however, indicators of changes in service demand and supply, city government form, and fire-fighter as well as non-fire-fighter unionization must be specified and taken into account. In general, we should expect that as demand for fire protection in a city increases, that wages should also go up. Conversely, as the supply of potential fire-fighter applicants increases, wages should be lower. The demand for fire protection is a function of a city's need or taste for the service as well as its ability to pay for the service. As a city's need for fire-protection services increases, demand should increase and be associated with higher wages and lower working hours. However, even though the need or taste might be high, inability to pay for the services would dampen demand. Thus the lower the increase in ability to pay, the lower should be wages and the higher should be working hours.

As with demand, fire-fighter supply is posited to be a function of several factors. First, the supply of fire fighters should be a function of the wage potential of the occupation. If other occupations requiring similar skills pay higher wages over time in the same city, the pool of applicants should decrease. This low supply should be conducive to higher wages and lower working hours. Holding the wage potential constant, we would expect several additional factors to be of import. A second factor affecting labor supply is the level of education of city residents. As individual educational levels increase, the pool of qualified applicants should be larger, thus depressing wages and increasing hours. Finally, the proportion of the population that is black should be inversely related to the pool of potential fire-fighter applicants. While this may be the result of either discrimination on the part of fire fighters or the low nonmonetary benefits that blacks may derive from fire protection, it should result in a small supply of potential applicants.[21]

Indicators of unionization must also be specified in our model. Generally, as unionization increases, wages should be higher and working hours lower. Several aspects of unionization in each city should be relevant, however. First, the extent of unionization should be important. Penetration of the employee group (fire fighters) by the union should be positively related to the union's ability to obtain demands. Because in general when an IAFF local exists it contains a large proportion of all fire fighters in the city, we may assume that the extent of unionization is high if a local exists.[22] Second, in cities where fire-fighter locals exist, union strength must be assessed. The greater the union's power, as indicated by the existence of written contracts and security provisions such as dues checkoff systems, the more likely it is that wages will increase and hours decrease.

In addition to the above mentioned supply-and-demand factors, another factor—city government form—may be of importance. City government form should be expected to affect the unionization-wage relationship for several reasons. First, as Ehrenberg suggested, city managers may be more efficient negotiators and stress efficiency least-cost notions more than elected officials.[23] Second, as a number of observers noted, the more diffuse decision-making structure as well as the ethos characteristic of mayor-council cities should produce a negotiation situation that is more fractionalized and offering more points of access and influence than is true for council-manager cities.[24]

Hypotheses

The hypotheses to be tested are:

1. Fire-fighter unionization will be positively related to fire-fighter wages and inversely related to working hours.

2. The demand for fire protection will be positively related to fire-fighter wages and inversely related to working hours.
3. Fire-fighter labor supply, however, should be inversely related to wages and positively related to working hours.
4. Mayor-council form of city government should be positively related to fire-fighter wages and inversely related to working hours. Council-manager form of city government, however, should be inversely related to wages and positively related to working hours.

Data and Analysis Technique

The following predictors are used in analyzing the wage and working hour impacts of fire-fighter unionization.[25–28]

Fire-Protection Demand
1. Municipal population density
2. Median value of housing in municipality

Fire-Fighter Supply
1. Average hourly earnings of manufacturing production workers
2. Median education level in municipality
3. Percentage of population black

Unionization
1. Existence of an IAFF local
2. IAFF contract with municipality
3. IAFF dues checkoff provision with municipality

City Government Form
1. Existence of council-mayor/council-manager government

All independent variables, with the exception of city government form, are "change" variables. In assessing the impact of changes in the existence of a fire-fighters local in each of the cities changes were computed for all variables for the 1960 to 1970 time period. Changes in the union contract and dues checkoff status of cities with a fire-fighter local, however, were available only for the 1966 to 1970 period. Thus the latter equations utilize 1960 to 1970 changes for the supply-and-demand variables, and 1966 to 1970 changes for the unionization and dependent variables.

Two dependent variables are utilized in the analysis. They are the annual entrance salary for fire fighters in each municipality, and the number of weekly work hours. Because we are interested in relating city government form, changes in demand-and-supply factors, and unionization, to changes in wages and hours, the dependent variables were created by residualizing the later time period of each with the corresponding earlier time period. Thus fire-fighter annual entrance salary and working hours were computed for 1960 and 1970 and the later periods regressed upon the earlier ones. The dependent variables are the residuals. We predict, as best we can, assuming incremental changes, and analyze the deviations from these predictions. In this case, we are comparing changes across cases, where we examine shifts in each dependent variable apart from what we would expect, given national trends. This nonincremental change is obviously "harder" to explain/predict, hence amounts of variance "explained" will be lower. However, the variance explained can more reliably be attributed to actual changes in the independent variables rather than false historical covariance between socioeconomic development factors and each dependent variable.

Because the predictor variables are either interval or dichotomous and the dependent variables are interval in nature, multiple regression analysis is used as the basic technique of analysis. Not only does this technique permit an assessment of the independent impact of each factor upon wages and hours, but the combined impact can be assessed as well.

Findings

Tables 16-1 through 16-4 present the results of the regression analyses. The impact of union local changes upon working hours and entrance salary is contained in tables 16-1 and 16-2, respectively. Union strength, as measured by union contracts and dues checkoff provisions, and its impact upon working hours and entrance salary is presented in tables 16-3 and 16-4.

A change to a union local during the 1960 to 1970 period is one of three factors having a statistically significant impact upon changes in weekly work hours (table 16-1). Most important is form of city government. In fact, fire fighters in city-manager cities work 2.7 hours per week longer than fire fighters in mayor-council and commission cities. As expected, the formation of a fire-fighters local during the time period reduces the work week by 2.1 hours. In general, supply-and-demand factors are relatively unimportant when compared to the city government form and unionization factors. Only one supply factor—percentage of the population that is black—has a sizable impact upon working hours, with an increase in the proportion of the population that is black reducing working hours, as predicted. As fire-fighter supply decreases, working hours are also decreased.

Union impacts upon salary are shown in table 16-2. A change to a fire-fighters local is again statistically significant in its impact. For cities adopting a union local during the 1960 to 1970 period, entrance salaries increased by $217 more than would be expected, given national trends.

Table 16-1
Regression Equations for Changes in Hours per Week for All Cities (1960-1970): Union Local

Variable	*b*	*Beta*	*Marg. R^2*
Demand change, 1960-70			
Density of pop.	- .0001	-.0251	.0005
Med. value housing	.0000	.0346	.0010
Supply change, 1960-70			
Med. education level	.0113	.0145	.0002
% black pop.	- .0932	-.1136[a]	.0108
Mfg. wages	.0039	.0147	.0002
Government form, 1968			
City-manager form	2.7025	.2542[a]	.0595
Unionism change, 1960-70			
Union formed	-2.1613	-.1762[a]	.0294
R^2 = 10.6%; a = -1.6008			

Note: Changes are residualized by regressing 1970 figures on 1960 values.

[a]$T > 2.00$ ($p < .05$); n = 411.

Table 16-2
Regression Equations for Changes in Salary for All Cities (1960-1970): Union Local

Variable	*b*	*Beta*	*Marg. R^2*
Demand change, 1960-70			
Density of pop.	.0095	.1507[a]	0.186
Med. value housing	.0012	.0083	.0001
Supply change, 1960-70			
Med. education level	- 12.5100	-.1103[a]	.0104
% black pop.	16.6865	.1395[a]	0.163
Mfg. wages	9.2933	.2396[a]	.0527
Government form, 1968			
City-manager form	54.0010	.0348	.0011
Unionism change, 1960-70			
Union formed	217.1292	.1214[a]	.0140
R^2 = 17.4%; a = -1.006.4.			

Note: Changes are residualized by regressing 1970 figures on 1960 values.

[a]$T > 2.00$ ($p < .05$); n = 411.

Unlike working hours, however, supply-and-demand factors are important in determining changes in entrance salary. Most important are changes in manufacturing wages in the community, an indicator of alternate wages, with increases leading to higher fire-fighter entrance salary. Likewise, decreases in labor supply, as indicated by the percentage of population that is black variable, and increases in demand, as indicated by the population-density variable, lead to higher-entrance salary. Also as expected, increases in the median education level of a municipality's population leads to decreases in entrance salary. This is most likely the result of a larger pool of qualified fire-fighter applicants, thus making for a more competitive labor market (for applicants) with resulting lower wages.

The impact of union strength upon working hours and entrance salary is presented in tables 16-3 and 16-4. As noted earlier, the change period for these two equations is 1966 to 1970. Thus for those cities with fire-fighter locals in 1966, we look at changes in union strength, as measured by contracts and dues checkoff provisions, and changes in working hours and entrance salary for the same time period.

The equation for working hours is contained in table 16-3. The unionization and city government form indicators are the most important indicators of changes in working hours. The existence of a city manager increases fire-fighter workweeks by 1.9 hours. Change to a dues checkoff provision, one of the two indicators of union strength, counteracts the impact of the city-manager form

Table 16-3
Regression Equations for Changes in Hours per Week for Cities with Unions in 1966: Union Contracts and Dues Checkoff

Variable	*b*	*Beta*	*Marg. R^2*
Demand change 1960–70			
Density of pop.	- .0004	-.0966	.0078
Med. value housing	.0000	-.0365	.0011
Supply change, 1960–70			
Med. education level	.0081	.0113	.0001
% black pop.	- .0753	-.0980	.0081
Mfg. wages	- .0053	-.0220	.0004
Government form, 1968			
City-manager form	1.8877	.1866[a]	.0332
Unionism change, 1966–70			
Union contract	- .8206	-.0715	.0045
Dues checkoff	-1.3709	-.1087[a]	.0105
R^2 = 8.5%; a = -.7423.			

Note: Changes are residualized by regressing 1970 figures on 1960 values.

[a]$T > 2.00$ ($p < .05$); n = 342.

of government by decreasing the workweek by 1.4 hours. Union contracts do not have a statistically significant impact upon working hours. Supply-and-demand factors have only a negligible or no impact upon fire-fighter working hours.

The impact of changes in union strength upon changes in entrance salary is contained in table 16–4. As can be seen, both of the union strength indicators have little or no impact upon salary changes. Supply-and-demand factors are once again most important in predicting change in fire-fighter entrance salary.

Most important are changes in manufacturing wages, with increases in wages in the manufacturing sector leading to annual entrance-salary increases for fire fighters. Also having a statistically significant impact upon entrance salary are the other two supply indicators—changes in the percentage of the population that is black and median education level. One of the demand indicators (population density) is again an important predictor of entrance-salary increases. Of overwhelming importance, however, is the manufacturing wage indicator, accounting for almost 6 percent of the variance in entrance-salary changes itself.

Conclusions

In summary, we can see that a change to a union local has a significant impact upon fire-fighter working hours and entrance salary. The formation of a union

Table 16–4
Regression Equations for Changes in Entrance Salary for Cities with Unions in 1966: Union Contracts and Dues Checkoff

Variable	*b*	*Beta*	*Marg. R^2*
Demand change, 1960–70			
Density of pop.	.1067	.1590	.0213
Med. value housing	.0021	-.0138	.0002
Supply change, 1960–70			
Med. education level	-15.4985	-.1362[a]	.0166
% black pop.	14.9942	.1225[a]	.0126
Mfg. wages	9.8315	.2554[a]	.0588
Government form, 1968			
City-manager form	- 1.8067	.0011	.0000
Unionism change, 1966–70			
Union contract	63.7592	-.0349	.0000
Dues checkoff	4.1529	-.0021	.0012
R^2 = 15.8%; a = -1043.3.			

Note: Changes are residualized by regressing 1970 figures on 1960 values.

[a] $T > 2.00$ ($p < .05$); n = 342.

local decreases the workweek by 2.1 hours, while increasing wages at the same time by $217. While supply-and-demand factors are more critical in determining nonincremental entrance salary changes, the union local still has a significant impact.

To ascertain the degree to which a union was able to obtain further demands after it was formed, we looked at changes in union strength and changes in working hours and entrance salary. Again we found that unions had their greatest impact upon working hours, with dues checkoff provisions reducing the workweek by 1.4 hours from 1966 to 1970. Contracts were not found to have a statistically significant impact, though a possible trend was indicated by the regression equation. Finally, changes in union strength were found to have little or no impact upon changes in entrance salary, with the supply-and-demand factors being most important in contributing to changes in salary during the 1966 to 1970 period.

Unlike previous analyses we have found the unionization of fire fighters to have a moderate impact upon wages and working hours. Also while Ehrenberg found that the existence of a contract had a significant impact upon wages and hours, we found that the most critical of the two measures of unionization is the formation of a union local. It is likely that Ehrenberg's analysis may have understated the union impact upon wages and working hours since it was only a single "snapshot" of what are complex, dynamic, and often slow-moving processes. Our model, which is longitudinal in nature, allows an examination of these processes over a longer period of time.

In conclusion, the impact of unionization upon fire-fighter working hours and entrance salary should not be underplayed. Municipalities undergoing unionization should learn from this evidence and be prepared to bargain with municipal employees effectively. The fact that city-manager cities, where administrative decision making is often less diffuse and ambiguous, are effective in keeping working hours from being reduced by unions tends to support this thesis. Once unionization is established, however, it appears that factors that are often out of the control of municipal officials, such as the prevailing manufacturing wage in the municipality, are of greater import in determining employee wages.

Notes

1. For a fairly traditional overview of these arguments see O. Glenn Stahl, *Public Personnel Administration,* 7th ed. (New York: Harper and Row, Publishers, 1976), pp. 324–337. Also see David Lewin, "Public Employment Relations: Confronting the Issues," *Industrial Relations* 12 (1973): 309–321.

2. For a general statement of this nature see Frederick O'R. Hayes, "Collective Bargaining and the Budget Director," in Sam Zagoria, ed., *Public*

Workers and Public Unions (Englewood Cliffs, N.J.: Prentice-Hall, 1972), pp. 89-100.

3. For a brief overview of this literature see the review by David Lewin, "Public Sector Labor Relations," *Labor History* 18 (1977): 133-144.

4. This is even smaller than the average union wage impact in the private sector. See H. Gregg Lewis, *Unionism and Relative Wages in the United States* (Chicago: University of Chicago Press, 1963). Also see Michael Boskin, "Unions and Real Wages," *American Economic Review* 62 (1972): 466–472.

5. In 1960 there were 676 cities with a population over 25,000. We were able to collect unionization data for 419 cities. Thus our analysis is based upon data for around 60 percent of all cities over 25,000 population in 1960. Examination of differences between the study cities and all cities over 25,000 revealed no substantial differences for city size, region, and city governmental form.

6. See Hugh O'Neill, "The Growth of Municipal Employee Unions," in Robert H. Connery and William V. Farr, eds., *Unionization of Municipal Employees* (New York: Academy of Political Science, Columbia University, 1970), pp. 3–4.

7. Ibid., p. 4.

8. Jack Stieber, *Public Employee Unionism: Structure, Growth, Policy* (Washington, D.C.: Brookings Institution, 1973), p. 52.

9. The largest all-public employee union, of course, is the American Federation of State, County, and Municipal Employees (AFSCME), with the second largest union being the American Federation of Teachers (AFT).

10. Stieber, *Public Employee Unionism,* p. 8.

11. Data indicating this can be found in Jack Stieber, "Employee Representation in Municipal Government," *The Municipal Year Book, 1969* (Washington, D.C.: International City Management Association, 1969), pp. 31-57; and Boyd Hartley, "Fire Department Unionization," *The Municipal Year Book, 1970* (Washington, D.C.: International City Management Association, 1970), pp. 390–404.

12. Orley Ashenfelter, "The Effect of Unionization on Wages in the Public Sector: The Case of Fire Fighters," *Industrial and Labor Relations Review* 24 (1971): 192.

13. Sterling Spero, *Government as Employer* (New York: Remsen Press, 1948), pp. 228-244.

14. See James A. Craft, "Fire Fighter Strategy in Wage Negotiations," *The Quarterly Review of Economics and Business* 11 (1971): 67-76.

15. Stieber, *Public Employee Unionism,* pp. 204-207.

16. The most recent assessments are Roger W. Schmenner, "The Determination of Municipal Employee Wages," *Review of Economics and Statistics* 60 (1973): 83-90; James L. Freund, "Market and Union Influences on Municipal Wages," *Industrial and Labor Relations Review* 27 (1974): 391–404; Donald

E. Frey, "Wage Determination in Public Schools and the Effects of Unionization," in Daniel S. Hamermesh, ed., *Labor in the Public and Nonprofit Sectors* (Princeton, N.J.: Princeton University Press, 1975), pp. 183-219; Daniel S. Hamermesh, "The Effect of Government Ownership on Union Wages," in Hamermesh, *Labor in the Public and Nonprofit Sectors,* pp. 227-255; Ronald G. Ehrenberg and Gerald S. Goldstein, "A Model of Public Sector Wage Determination," *Journal of Urban Economics* 2 (1975): 233-245; W. Clayton Hall and Norman E. Carroll, "The Effect of Teachers Organizations on Salaries and Class Size," *Industrial and Labor Relations Review* 12 (1973): 834-841; and Hirschel Kasper, "The Effects of Collective Bargaining on Public School Teacher's Salaries," *Industrial and Labor Relations Review* 24 (1970); 57--72.

17. Ashenfelter, Case of Fire Fighters," pp. 191-202.

18. Ibid., 201.

19. Ronald G. Ehrenberg, "Municipal Government Structure, Unionization, and the Wages of Fire Fighters," *Industrial and Labor Relations Review* 27 (1973): 36-48.

20. Ibid., pp. 47-48.

21. This is also argued by Ehrenberg, though the most likely reason for the low number of black fire fighters is discrimination. For more on fire fighters and blacks see Stieber, *Public Employee Unionism,* pp. 63-66.

22. See note 12 for data buttressing this argument.

23. Ehrenberg, "Wages of Fire Fighters," p. 40.

24. See, for instance, Thomas M. Love and George T. Sulzner, "Political Implications of Public Employee Bargaining," *Industrial Relations* 11 (1972): 21-22.

25. These data are from the 1962 and 1972 editions of the *City and County Data Book.*

26. Ibid.

27. Data for the unionization indicators are taken from the 1960, 1961, 1966, and 1970 editions of the *Municipal Year Book.*

28. Data for this variable are drawn from the *Municipal Year Book,* 1968.

17 Representation in the U.S. Armed Forces?

Ezra S. Krendel and
Bernard L. Samoff

On November 8, 1978, President Carter signed bill S. 274 which then became Public Law 95-610. This law prohibits union organization of the armed forces, membership in military labor organizations by members of the armed forces, and recognition of military labor organizations by the government. Should unions enroll military personnel in order to engage in collective bargaining, they could be fined up to $250,000 under this law. Military personnel who knowingly join a military union or solicit members for it are subject to a fine of up to $10,000 and as much as five years in jail. This law follows two years of legislative activity whose final result was delayed by two factors.

First, the Department of Defense (DOD) did not feel that S. 274 and similar bills were necessary since existing regulations, DOD Directive no. 1325.6, 12 September 1969, and DOD Directive no. 1354.1, 6 October 1977, were adequate to the problem.[1] In the more recent directive prohibited activities include collective bargaining, strikes, picketing to induce or coerce other service personnel to strike or slow down, recruitment for unions on military installations, and membership in a military union when such organizations present "a clear danger to discipline, loyalty, or obedience to lawful orders." However, the directive does not prohibit presentation of grievances through established channels nor bar commanders from considering the views of command-sponsored or authorized advisory councils. Nor does it prohibit individual or collective petitions to or communications with members of Congress. Moreover, a service member may join a union in connection with off-duty employment, and a civilian working at a military installation may join a union to represent him with respect to terms and conditions of employment. Secretary of Defense Harold Brown preferred regulations to legislation because the latter would be more vulnerable to challenge under the First Amendment.[2]

The second factor in delaying P.L. 95-610 was the strong difference between the bill as passed by the Senate and the more than twenty similar bills passed by the House of Representatives. The original Senate version of S. 274 differed significantly from these bills in that it additionally prohibited union membership by civilian technicians employed by reserve and guard units who are also military members of those units.[3] Most technicians are represented by unions as allowed under Executive Order 11491.

The House deleted the prohibition on union membership by civilian technicians, and in the closing days of the 95th Congress, the Senate agreed. The bill as passed, providing for sanctions making union activity a criminal offense, is substantially the same as the current DOD directive.

P.L. 95-610 and the referenced DOD directive were sparked by efforts, particularly those of the American Federation of Government Employees (AFGE), to organize the armed forces. Within the constraints of this study, we will not describe these efforts except to note they arose out of a feeling on the part of AFGE leadership that their efforts to maintain pay comparability with the private sector for their membership were, because of the statutory linkage between pay for the armed forces and the civilian pay scale in the federal service, also of clear financial benefit to military personnel. From this point of view, personnel in the armed forces were getting a free ride on AFGE efforts. The armed forces personnel, however, exhibited no reluctance to accept these advantages at no cost. When AFGE surfaced the idea of organizing the armed forces this stimulated a variety of work-related and enlistment "contract" complaints from military personnel, which were interpreted by AFGE as enthusiasm for military unionization.

On September 7, 1977, AFGE announced it was abandoning plans to organize the military. Citing a four-to-one opposition vote by AFGE locals, President Ken Blaylock reproted that AFGE would not organize GIs now or at any time in the foreseeable future. The decision is a reversal of the actions at the 1976 AFGE convention when Blaylock won the constitutional authority to bring uniformed service personnel into the union. No union is currently challenging P.L. 95-610 by attempting to organize the armed forces, though the size of potential membership and the dissatisfactions of service personnel, however weakly indicated, may eventually induce one or more unions to provoke a court test.

At this stage—restrictive legislation and regulations and no active organizing—we can fruitfully consider three foci with implications for policies directed toward maintaining tranquil personnel relations in the armed forces: (1) the changing armed forces; (2) DOD management response to the changing armed forces; and (3) organizations of armed forces personnel.

The Changing Armed Forces

For various reasons, many arising out of experience with the warn in Vietnam, conscription ended in June 1978 and the All Volunteer Force (AVF) came into being.[4] This change from a large conscript armed forces to a nearly equally large peacetime volunteer force, unquestionably one of the most significant in the peacetime history of the armed forces, was accompanied by substantial increases in salaries and benefits. By design and action in appealing to economic forces, it heralded the emergence of military service as an occupational role, and brought into closer alignment public/private industrial/service workers and armed forces personnel.

The AVF impelled DOD to go into the labor market to obtain conscripts by stressing comparable pay, benefits, training and careers, the physical surroundings provided personnel, choice of station and occupation (for example, The Army Wants to Join You!), and the due process safeguards. To the extent that the forgoing were the marketing pitch of DOD, appeals to duty, honor, and patriotism were scarcely mentioned. The shift to AVF implies a contractual basis for military employment, with the accompanying decline in the notion of service as a duty. We can expect an increasing number of suits in both the military and civil courts over the issue of claimed violations of enlistment contracts in such matters as the geographical region to which the service person was assigned, the nature of quarters, whether promised training programs came up to expectations, and the extent to which benefits existing at enlistment are maintained.

An important inducement for military enlistment is the desire for vocational training and economic advancement. Potential recruits cite job training and skill development as the key advantages of military service.[5] In the context of this market economy it is no surprise that many AVF service personnel tend to view themselves as employees. This view is further reinforced by recruiting advertising which leaves the impression that military life and working conditions are or can be similar to those in the civilian world. When actual conditions compare poorly with the recruitment promises, the serviceman becomes dissatisfied and becomes a likely candidate for representation.

It is hardly necessary to emphasize the implications for representation of the contrasts between the occupational model and the calling or professional one. As the former gains ascendancy, armed forces personnel become civilianized and their tasks, particularly as these become more specialized and technical, become more like those of the civilian labor force. These dynamics seem ineluctable and imply that irrespective of traditional and even legal obstacles, the armed forces will be pressured into exerting authority and discipline in a manner more compatible with civilian values. This is not surprising since over the centuries the armed forces modified their allowable exercise of authority to reflect the values of the nation as a whole. Flogging has long been outlawed in the U.S. Navy.

There is also evidence that the management group of the armed forces, senior noncommissioned officers and officers, has been dissatisfied because of a retrenchment in economic benefits, a perceived violation of enlistment contracts, and a limiting of career opportunities. Specifically, there are complaints about forced separations, pay limitation, reduction in medical benefits, and challenges to retirement pay. Part of the problem lies in the peacetime nature of the armed forces with inevitable reduction in career opportunities. Another factor is the current economic crisis and government efforts to limit rising federal expenditures.

Surveys designed to assess the attitudes of armed forces personnel to military unions are deficient in at least two major respects. First, respondents do not understand the implications of such unions and, second, the surveys occur in the absence of any specific circumstances intensifying discontents and stimulating unionization. Be this as it may, however, two recent surveys carried out with the permission of but not under the direction of the military command, one of U.S. Air Force personnel and the other of U.S. Army combat arms personnel, both indicated that about 30 percent of the enlisted respondents favored a union; the officers were overwhelmingly opposed.[6] Such surveys probably indicate a diffuse dissatisfaction with the individual's weakness in the face of authority as well as with the manner by which this authority is exercised, more than a readiness to form a union.

DOD is seeking to reduce manpower costs. Expenditures for weapons' systems are rising; to operate within budgetary constraints and maintain an adequate defense capability manpower costs and benefits must be reduced.[7] To achieve this DOD is trying to limit pay increases, restructure military retirement, eliminate commissary subsidies, reduce the medical program for dependents, change the allowance for quarters, and limit terminal leave payments. The armed forces are reacting adversely to these cutbacks, forced retirement and reductions-in-force. Representation thrives under these conditions.

The armed forces, however, are under pressure to change once again. Bills have been prepared for submission in the current Congress to reinstitute registration of men and women between the ages of 18 and 26 so as to facilitate conscription in the event of a national emergency. Although there is some sentiment for a relatively expensive program of compulsory service by the registrants in either the civilian or the military sector, it appears more likely that should registration be instituted, military conscription will be implemented concurrently or will follow rapidly. Civilian national service would create unacceptable budgetary demands. Should peacetime conscription come about once again, it is highly likely that a large variety of cultural and interpersonal conflicts will erupt between conscripts who presumably will be representative of a broad range of American socioeconomic classes and the cadre of "lifers" who in the main have their roots in the economically disadvantaged classes. Efforts to achieve collective representation may be expected to thrive under these conditions too.

DOD Response to the Changing Armed Forces

The reality of the AVF and collective action has both catalytic and positive effects upon management. Before discussing these we should distinguish between the chain of command and management. Experience with private-sector bargaining, where the analog to command is unchecked hierarchical authority, and public-sector bargaining in essential services (police officers, fire fighters, and prison guards) reveal no serious infringement upon the line authority of command to direct employees during operations. Collective representation, as such, does not interfere with the chain of command. Just as these employees use self-help, such as blue-collar flue, sick-outs, slowdowns, and working by the book, solely to improve their wages, hours and working conditions, and not to challenge command, so we would expect armed forces personnel to act in a similar manner. These tactics are used because picketing and striking are unlawful.

More important, collective representation has two major thrusts. First, it lends a consensual legitimacy, with defined limits of joint participation, to the superior's authority. In this setting orders must be functionally relevant to the tasks required by the organization's structure, not capricious, arbitrary, or emanating from the incumbent's position alone.

Second, it impels management to rationalize and to improve their performance. This occurs because of a continuing feedback from employees via grievances and of the need to compromise a large variety of interests. In the absence of formal representation, management lacks effective channels for assessing its performance. In formulating policies and implementing them, management must balance all interests, including those expressed by employees.[8]

If the armed forces change and acquire more of the attributes of public-sector employment, some form of representation will evolve and inescapably tensions with command and management will develop. These tensions need not be counterproductive for the military. If addressed with imagination and skill and if the arena of disagreement is clearly limited to living-condition associated issues, they may in fact lead to a more responsive and effective armed forces.

Conflicts over discipline could diminish through a grievance procedure; due process for grievances not covered by the Uniformed Code of Military Justice is important to the armed forces. We do not believe that formal inputs from military personnel on such items as medical benefits, hair length, transfers, promotions, or location allowance will impair command authority and discipline. Formal inputs, of course, are routinely made over compensation issues when the decision authority is Congress rather than the management of the armed forces. As evidence, witness the effectiveness of the various service personnel and veterans organizations in lobbying for pay, benefits, and preferential treatment for their constituents.

On balance, and without minimizing the frictions and differences, genuine representation in conferring with management could improve the armed forces. It could stimulate management to develop rational and acceptable rules; eliminate trivial, annoying, and unproductive rules and practices; and consider the needs and demands of the armed forces more sympathetically. Once unfettered command power over employment matters gives way in specific areas to collective relationships, management must think about managing, exploit knowledge about human resources, search for accommodation with the independent power of service personnel, and share certain decisions with their representatives. Such a development in the management of the armed forces is particularly indicated should military conscription become law once again.

Organization of Armed Forces Personnel

There are at least twenty-two organizations which lobby, with varying degrees of intensity and effectiveness, for the interests of armed forces personnel, past and present.[9] The American Legion, a powerful interest group with nearly 3 million members, adopted a resolution at its national convention in 1975 in Minneapolis which states in part: "Armed Forces personnel are adequately represented by senior Armed Forces personnel, their elected representatives . . . and by organizations of interested and influential U.S. Armed Forces veterans, such as the American Legion."

Three hundred delegates to the Air Force Sergeants Association meeting in Las Vegas, Nevada, in early August 1975 passed resolutions denouncing the "erosion of promised enlistment benefits and the threat of further reduction in the retirement system" In a similar vein, 405 delegates to the Fleet Reserve Association's (FRA) forty-eighth National Convention in St. Louis, Missouri, in September 1975 heard a report from a special national committee. One of its findings was that FRA has provided benefits and services of a union without the threat of eroding the military command structure. The delegates also instructed the national committee on legislative service to pursue all legislative goals and beneficial personnel policies with decisive action and extra vigor to demonstrate to all military personnel that adequate representation can be provided without a union.

The above illustrates both the representational character of the organizations working on behalf of the military personnel and their concern that unionization may encroach upon their territory. Even though these organizations oppose unionization and currently lack the attributes of a labor organization as defined in our national labor laws, they will continue acting in a representational capacity and may evolve into labor unions, as did comparable organizations in education and as some are in the process of doing in engineering.

For example, in the teaching profession the National Education Association (NEA), a nationwide organization of teachers, principals, and administrators, concentrated in its early years on professional and technical matters. As economic and other conditions changed, the NEA was prodded into a union stance by the American Federation of Teachers (AFT), an avowed teachers' union from its inception. Modifying its internal structure and excluding from membership supervisors and nonteaching administrative educational employees, the NEA embarked on an aggressive drive of union activity among teachers in competition with AFT. The most recent figures reveal that NEA has a membership of 1,470,000 compared to AFT's membership of 444,000.[10]

Of course, we cannot predict that existing organizations of current and past armed forces personnel will evolve into unions. Just as the NEA evolved into a union, though still retaining many characteristics of a professional organization, the Air Force Sergeants Association and other organizations of military professionals could be impelled into the posture of military unions, if only to remain viable to their members. Like most boundaries, that between unions and unionlike organizations is hard to draw with precision. These existing twenty-two or more organizations take an active interest in the economic status of their members and in the posture of the military in American society. Such transformations to unionlike representation is quite feasible in view of the experience of NEA and other professional organizations, such as the American Association of University Professors (AAUP).

No one disputes the manifest and latent powers of these twenty-two plus organizations. Indeed, they obtained approval in the House of Representatives by a 327 to 7 vote to retain veterans' preference provisions in hiring and retention for federal jobs in the Civil Service Reform bill.[11] Considering their power, organization, and resources, we would expect them to change to collective, unionlike representatives. The history of American unions suggests such transformations; structures are adapted to meet needs, functions, and purposes.

A possible route for shifting to collective representation may be found in both P.L. 95-610, section 2(3) and section F2 of DOD Secretary Harold Brown's October 6, 1977, directive which states:

> F. PERMISSIBLE ACTIVITY
> This Directive does not prevent, among other things:
>
> 2. Commanders or supervisors from giving due consideration to the views of any member of the armed forces presented individually or as a result of participation command-sponsored or authorized advisory councils, committees or organizations for the purpose of improving conditions or communications at the military installation involved.

Whichever of the twenty-two plus organizations qualify as "command-sponsored or authorized advisory councils, committees or organizations," they would communicate with commanders regarding improved conditions, due process procedures, and complaints of members of the armed forces. Although this falls short of conventional bilateral collective bargaining, it does provide access to authority and is a developmental stage to collective representation.

Policy Implications

The armed forces have changed and are still in the process of change. If the AVF fails and yields to a combination of volunteers and conscripted personnel, the armed forces may subsume a cadre of lifers in the occupational mode and consequently susceptible to some of the appeals of representation. Concurrently, the potential for interpersonal clashes between this cadre and the conscriptees is such as to cry out for some form of institutionalized representation to handle a wide range of expected grievances. Such representation needs to be effective to satisfy the assertive and comparatively sophisticated youth from the general population in the 1980s. Absent such representation, a variety of ad hoc disruptive and politically motivated groups vying to represent the conscripts may be expected to arise.

It thus appears that however the changes in the armed forces proceed and despite the strictures of P.L. 95-610, some form of representation will arise in the military. The challenge to the managers of the armed forces is to act early on so that representation becomes a beneficial force promoting an effective American military which is compatible with our attitudes and values as we enter the twenty-first century.

Notes

1. Ezra S. Krendel and Bernard L. Samoff, *Unionizing the Armed Forces* (Philadelphia: University of Pennsylvania Press, 1978), pp. 3-6.

2. Daniel P. Sullivan, "Soldiers in Unions–Protected First Amendment Right?" *Labor Law Journal* 20 (September 1969): 581-589; Paul D. Staudohar, "Legal and Constitutional Issues Raised by Organization of the Military," *LLL* 28 (March 1977): 182-186; James K. McCollum and Jerald F. Robinson, "The Law and Current Status of Unions in the Military Establishment," *LLJ* 28 (July 1977): 421-430.

3. The evolution of this bill can be seen from S. 3079, 94th Cong. 2d sess., and S. 274, 95th Cong., 1st sess., both as introduced and passed by the Senate, and as amended by the House Committee on Armed Services. The House bills are H.R. 51, 120, 384, 624, 675, 693, 1105, 1381, 1478, 1623, 2230, 2477, 2478, 2479, 2678, 2926, 2983, 3069, 3262, 3271, 3524, 4040, 4927, 5008, 5139, 8245, and 8326.

4. *The Report of the President's Commission on an All-Volunteer Armed Forces* (Washington: U.S. Government Printing Office, 1970).

5. Opinion Research Corporation, *Attitudes and Motivations toward Enlistment in the U.S. Army* (Princeton, N.J.: University of Princeton Press, 1975), p. 17.

6. T. Roger Manley, Charles W. McNichols, and G.C. Saul Young, "Attitudes of Active Duty U.S. Air Force Personnel toward Military Unionization," *Armed Forces and Society*, vol. 3, no. 4 (Summer 1977); David R. Segal and Robert C. Kramer, "Ground Forces Attitudes," in *Military Unions: U.S. Trends*

and Issues, eds., William J. Taylor, Jr., Rogert J. Arango, and Robert C. Lockwood (Beverly Hills, Calif.: Sage Publications, 1977).

7. U.S. Congress, Senate, Subcommittee on Defense of the Appropriations Committee, *Hearings,* 94th Cong. 2d sess., 1976, vol. 2, p. 15.

8. "The challenge that unions presented to management has, if viewed broadly, created superior and better balanced management. . ." Sumner H. Slichter, James J. Healy, and E. Robert Livernash, *The Impact of Collective Bargaining on Management* (Washington, D.C.: Brookings Institution, 1960), p. 951.

9. Krendel and Samoff, *Armed Forces,* pp. 9-11.

10. Appendix D, pp. 101-102, *Directory of National Unions and Employee Associations,* 1975, U.S. Department of Labor, Bureau of Labor Statistics, 1977, Bulletin 1937.

11. *The Philadelphia Inquirer,* September 14, 1978, p. 3-A.

Index of Names

Index of Subjects

About the Contributors

Edwin M. Epstein is professor of business administration at the University of California, Berkeley, and was an associate dean of the School of Business. Much of his published research has examined the role of business in American politics and the interconnection of government and business in the United States. In 1977-1978 he was a Fellow at the Woodrow Wilson International Center for Scholars in Washington, D.C., where much of the research for his chapter in this book was conducted.

J. David Gillespie received the Ph.D. from Kent State University. He is an associate professor of political science at Presbyterian College. Most of his published research has been in American politics, race relations, constitutional law, and international politics.

Helen Ginsburg, Ph.D., is associate professor of economics at Brooklyn College, City University of New York. She specializes in labor, social welfare, poverty, and urban economics. Among her publications are *Poverty, Economics and Society* and *Unemployment, Subemployment and Public Policy.* Dr. Ginsburg is also author of numerous articles on unemployment and full employment and has presented testimony on those topics before various congressional committees. Presently, she is engaged in research on Swedish labor-market policies for full employment.

Michele Hoyman received the Ph.D. in political science from the University of Michigan. She is assistant professor of industrial relations at the Institute of Labor and Industrial Relations at the University of Illinois at Urbana-Champaign. Her research interests include labor-union compliance with title VII of the Civil Rights Act, lobbying efforts for the Common Situs Picketing Bill, the consent decree as a civil-rights compliance mechanism, labor-union women in politics, and black participation in local unions.

Leon H. Keyserling is a graduate of Columbia University and of Harvard University, where he received the LL.B. degree. He has served as chairman of the President's Council of Economic Advisers under President Truman and as a consulting economist and attorney, and he is president of the Conference on Economic Progress.

Ezra S. Krendel, professor of operations research at The Wharton School, University of Pennsylvania, is known for his pioneering research in the mathematical description of human control behavior. Over the past twenty-five years he has participated in and been consulted on a variety of planning and cost-effectiveness studies for the U.S. Departments of Defense, Commerce, Transportation, and Justice, as well as for industries and municipal government.

Charles M. Lamb is assistant professor of political science at the State University of New York at Buffalo. He has served on the staff of the U.S. Commission on Civil Rights in Washington, D.C., and as a research scientist at George Washington University's Program of Policy Studies. Dr. Lamb is the author of several law review articles that address civil rights issues and judicial politics.

William Lyons is an associate professor in the Department of Political Science and associate director of the Bureau of Public Administration at the University of Tennessee, Knoxville. His articles dealing with urban politics and policy have been published in *American Journal of Political Science, Journal of Politics, Political Methodology,* and *Social Science Quarterly.*

Mary K. Marvel is assistant professor of public administration at The Ohio State University. Her research interests include program implementation and evaluation.

Robert A. McLean received the Ph.D. in industrial and labor relations from Cornell University. He is an assistant professor of business at the University of Kansas. He previously held positions at Hobart and William Smith Colleges, the University of Wisconsin–Milwaukee, and the American Medical Association.

Michael L. Mitchell received the B.A. from Samford University and the M.A. from Kent State University. He is a doctoral candidate and Teaching Fellow in the Kent State Department of Political Science. He has previously presented and published papers in American politics and in the teaching of political science.

Laura Katz Olson is assistant professor of government at Lehigh University. Her research has been in public-policy analysis with a focus on aging and retirement. She has been the recipient of a Fulbright-Hays Award, Gerontological Fellowship, and Mellon Faculty Development Grant. She serves on the Advisory Committee of the Pennsylvania Task Force on Aging and is coordinator of the Women's Studies Program at Lehigh. Her recent publications have appeared in *Polity, The Social Science Journal,* and *Policy Studies Journal.*

Robert C. Rodgers is a Ph.D. candidate in the School of Labor and Industrial Relations, Michigan State University. His research focuses on the effects of state-level collective-bargaining legislation on public-sector strike activity. Mr. Rodgers was formerly an assistant county administrator in Virginia.

Bernard L. Samoff is associate chairman and adjunct professor of the Department of Management at The Wharton School, University of Pennsylvania. His areas of expertise include industrial relations, labor law, and public administration. He is both an arbitrator and a fact finder, and has served as chief field examiner and regional director for the National Labor Relations Board.

Harvey L. Schantz received the Ph.D. from The Johns Hopkins University. He is assistant professor of political science at The State University of New York, Plattsburgh. Dr. Schantz served as an American Political Science Association Congressional Fellow during 1977-1978.

Richard H. Schmidt received the LL.B. from The George Washington University Law School. He is assistant counsel for legislation in the Division of Legislation and legal counsel of the Office of the Solicitor, U.S. Department of Labor. He served as an American Political Science Association Congressional Fellow during 1977-1978.

Ronald G. Schneck received the B.A. degree in political science and the M.A. in economics from the University of Wisconsin–Milwaukee.

James H. Seroka is assistant professor of political science at Southern Illinois University, Carbondale. His research interests include local-level policymaking, East European politics, and rural public administration. He is the author of recent articles in such journals as *Policy and Politics, Legislative Studies Quarterly, Slavic Review,* and *East European Quarterly.*

Russell L. Smith is an assistant professor in the Center for Governmental Studies, Northern Illinois University. His research interests are public-employee unionism, public management, and policy implementation. He has published articles in *Industrial and Labor Relations Review, Administration and Society, Midwest Review of Public Administration,* and *Policy Sciences Journal.*

Jeffrey D. Straussman is associate professor of public administration at the Maxwell School, Syracuse University, where he teaches courses in public budgeting and metropolitan government and politics. He is the author of *The Limits of Technocratic Politics,* and his current research focuses on public-sector responses to fiscal scarcity.

Joel A. Thompson is assistant professor of political science and director of research of the Appalachian Regional Bureau of Government at Appalachian State University. He received the Ph.D. from the University of Kentucky. His present research interests are in state and local policymaking, especially workmen's compensation, education, and voting rights.

Kenneth W. Tolo is associate vice president for academic affairs and associate professor of public affairs, The University of Texas at Austin. During 1978-1979 he directed the Apprenticeship Project at the Lyndon B. Johnson School of Public Affairs. His primary policy-research interests include employment and training programs, private pension-plan insurance programs, and higher-education management.

About the Editors

Charles Bulmer is an associate professor of political science at the University of Alabama in Birmingham. He received the Ph.D. degree from the University of Tennessee in Knoxville. His teaching interests include American government and public policy, normative political theory, and international relations. His articles on labor policy, international relations, and foreign policy have appeared in numerous professional journals. Dr. Bulmer is a member of Pi Sigma Alpha, the American Political Science Association, the Southern Political Science Association, the American Society of International Law, and the Policy Studies Organization.

John L. Carmichael, Jr., is an associate professor of political science at the University of Alabama in Birmingham, where his teaching interests include public policy, constitutional law, and American foreign policy. He received the J.D. degree from George Washington University and the Ph.D. degree from the University of Alabama. He is engaged in research on the entry of Great Britain into the European community. His publications include articles on labor policy, constitutional law, and foreign policy. He is a member of the Policy Studies Organization, the Southern Political Science Association, and the American Political Science Association.